D

The Robot Builder's Cookbook

The
Robot Builder's Cookbook

Owen Bishop

AMSTERDAM • BOSTON • HEIDELBERG • LONDON • NEW YORK • OXFORD
PARIS • SAN DIEGO • SAN FRANCISCO • SINGAPORE • SYDNEY • TOKYO
Newnes is an imprint of Elsevier

Newnes is an imprint of Elsevier
Linacre House, Jordan Hill, Oxford OX2 8DP
30 Corporate Drive, Burlington, MA 01801

Please Note: Although every care has been taken with the production of this book to ensure that the
projects contained herein operate in a correct and safe manner, the Publishers do not accept
responsibility for the failure of any project to work correctly or for any damage to any other
equipment it is connected to or used in conjunction with, or in respect of any other damage or injury
that may be so caused. The Publishers do not accept responsibility in any way for the failure to obtain
specified components.

Notice is also given that if equipment that is still under warranty is modified in any way or used or
connected to home-built equipment then that warranty may be void.

British Library Cataloguing in Publication Data
A catalogue record for this book is available from the British Library

Library of Congress Cataloging-in-Publication Data
A catalog record of this book is available from the Library of Congress

ISBN: 978-0-7506-6556-8

For information on all Newnes publications
visit our web site at www.newnespress.com

Printed and bound in UK

08   09   10   10   9   8   7   6   5   4   3   2   1

Contents

About this book

This is a book of practical robotics written for beginners but also catering for those who have progressed a little further beyond that stage. It describes the mechanics of robot construction, how to build the electronic circuits, and finally goes into the details of programming robotic systems.

The first half of the book is a cookbook of information, ideas, tips, and suggestions for the first-time roboticists and others. Much of the content will be of interest and practical use to students in Further and Higher Education who are working on a micro-controller-based project (though not necessarily a robotic one).

The second half of the book describes the designing, building and programming of five robots of varying degrees of complexity. The specifications are flexible and essential features are emphasised so that the designs are readily adaptable to whatever materials and parts the reader can obtain. Each description points the way to more advanced development of the project, resulting in a wide range of fascinating and often unique robots.

The programs are listed in the PICs MPASM assembler, which allows them to be modified, fine-tuned and extended. The listings are fully annotated and are accompanied by detailed flowcharts. These are intended to provide ample guidance for those who wish to program in one of the dialects of BASIC, or in the C language.

Companion website

This website carries downloadable files of the MPASM versions of all the programs and subroutines listed in the book. In addition there are files of programs for the *Quester* and the *Gantry* that are too long to be included in the book. All downloads are free of charge. The site also carries the same programs in the form of hexadecimal files.

The URL of the companion site is:

> http://books.elsevier.com/companions/9780750665568

Introduction

Making a Robot

Making a Robot

What sort?

The first question is — 'What sort of robot do we want to make?'.

When they hear the word 'robot', many people immediately think of the *R2-D2* or the robots of the film *I, Robot*. These are robots similar to humans in some ways, but not in all. There are many kinds of robot, one major group being the **mobile robots**, sometimes called **mobile platforms**. Examples of mobile robots include the human-like robots mentioned above and a wide range that mimic animals. Some walk about on six legs, like insects, and others jump around like frogs. Then there are the more useful mobile robots that run about the house, sweeping the floor, and those that find their way around a factory, delivering parts to the work-stations. These rarely look like humans — they just run around the place and *do* things.

Someone just starting in robotics, might begin with a low-cost mobile robot. Project 6.1, the *Scooter* (pp. 166-208), gives the details.

A mobile robot looks only vaguely human – and many do not look human at all!

Humans are capable of a wide range of tasks, but most robots are not so versatile. In industry, robots are designed to perform a very limited number of tasks, and to perform them precisely for hour after hour without getting tired or bored and without making mistakes.

In this category come the **robot arms** (opposite). These are usually not mobile. Robot arms are particularly useful for the heavy, unpleasant or repetitive tasks in industry. They can be used in environments in which it is harmful or dangerous for humans to work.

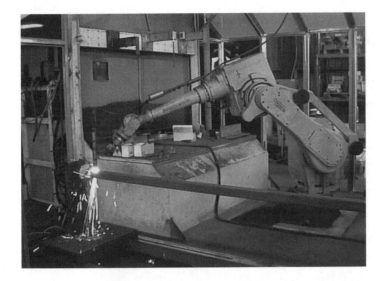

This robot has welding gear at the end of its arm. The arm is bolted to the floor but the welding torch can be manipulated so as to act at almost any location in the workshop. Other tools can be fitted to it when required.

Really heavy (in the sense of weight-lifting) tasks, need a **gantry robot**. Gantry robots are good at picking up massive items at one place and depositing them accurately at another place. Their main drawback is that they are not usually mobile, so the distance that they can transport the load is limited. In this book we use a gantry robot for light-weight tasks, such as picking up a playing-piece from a game board and moving it to the winning square. Our gantry robot needs brain rather than brawn.

So what shall it be? A mobile robot or a gantry robot?

Whichever the chosen type, the design process follows very much the same steps, as outlined in the remainder of this chapter.

This gantry robot at a printing works stacks up blocks of printed pages, ready for packing. As it stacks the blocks, it counts them and builds up a batch of definite size. Note one of its sturdy supporting columns in the foreground.

Getting down to detail

Having decided what type of robot to be built, the next step is to draw up a first specification. It may have to be revised later — at least its minor details.

Start the specification with a list of the things that robot should be able to do, its structure, and what electronic circuits it will use.

Refer to Parts 1 to 4 for ideas:

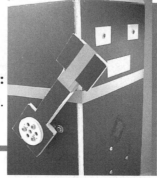

Part 1 Robot behaviour: what they can do.

Part 2 Robot mechanics: structures, materials, ready-made parts, tools.

Part 3 Robot electronics: components, sensors, actuators, and the circuits to drive them.

Parts 4 and 5 The PIC microcontrollers: How they interface to the robot electronics and how to program them to control the robot's actions.

This is a cookbook. There is no need to read Parts 2 to 5 from beginning to end. Just browse them to pick out any items of interest. Dip into these parts and gradually put together the final specification.

Controlling the robot

The first robots (they were called automatons in those days) were purely mechanical, driven by clockwork or steam power. The arrival of electronics greatly increased the scope of what robots could be made to do. Modern concepts of robotics began to emerge. The big advances came when engineers started putting complex digitial circuitry on a single chip. These were **microprocessors**, capable of millions of operations per second. Microprocessors are widely used in computers, robots, and many other devices that depend on high-speed, digital processing.

A microprocessor can not work on its own. There must be other electronic devices, such as memory chips, input and output ports, and a system clock, to help it. The circuit-boards holding the elements of a microprocessor system are relatively large and complex. They are just a bit too complicated for the average enthusiast to design and build.

Next came **microcontrollers** — the 'computer on a chip'. Fast operating, simple to connect to other electronics, easily programmable, and cheap to buy, they are just right for controlling simple robots.

A selection of microcontrollers. From left to right, the 18-pin PIC16F84, from Microchip, the 20-pin Atmel AT90S1200, and the 28-pin PIC16F87.
The data sheet in the background was down-loaded free from the Microchip site on the World Wide Web.

There are several manufacturers of microcontrollers (or controllers, as we will call them in future), but the PIC controllers made by Microchip Technology, Inc. seem to be the most popular in the hobby field (and in many professional fields, too). When you are planning to build a robot, it is important to choose the right PIC. Finding the right one is detailed on p. 130. The recently-introduced PIC16F690 is the one selected for the projects in Part 6, but several other types of PIC are equally suitable.

Programming a PIC

A controller operates according to a **program**. This is stored digitally in the controller's memory in the form of a code, called **machine code**. This code is very difficult to write by hand but, fortunately, a computer can help. Using special software, the program is typed in as a sequence of understandable instructions (or **mnemonics**) for the controller to execute. The software **assembles** the machine code from these instructions.

A PIC (top centre) being programmed in a programming deck. This is connected to a PC that is running special programming software. A display panel on the deck keeps the user informed of the progress of the operation.

Using a **programming deck** (photo opposite) the assembled machine code is copied from the PC into the memory of the controller. The deck usually has several light-emitting diodes (LEDs), and push-buttons for testing the output and input channels of the PIC while it is running the program on the deck.

The PIC recommended for the projects in this book have **flash memory**. Digital cameras use the same sort of memory for storing images. The advantage of flash memory is that it can be programmed and re-programmed over and over again, at least 100 times. So it is ideal for developing the software for a robot. Key in the program a section at a time, and test it as each section is completed. Later, parts of the program can be amended or deleted if something is wrong. Or even completely replaced with something entirely different.

As explained above, there is no need to write a program in the machine code in which it is eventually stored. Instead, the programmer writes in assembler. All PICs have the same assembler language, which has only 35 different instructions in it. This makes it quick to learn. Assembler is a one-action-per-instruction language. Step-by-step, the programmer tells the controller precisely what to do. Programs are easy to follow and understand.

Some people find assembler difficult because assembler instructs the controller in very small steps. They are not used to thinking in this way and prefer to program in bigger steps. They use one of a number of high-level languages. These include several dialects of BASIC, as well as C and C+. Instructions written in these languages are more like ordinary English. This makes programming easier, though it is still essential to pay strict attention to the syntax if the computer is to understand the program.

The machine code produced by the high-level language software (called a **compiler**) is usually appreciably longer than the code of an equivalent program written in assembler. It requires more memory to store it and the program does not perform as fast as an assembler program. This is not a problem for the robots described in this book, for the programs are short and high speed is not required.

If you wish to avoid any kind of programming, the machine code files are available on the Companion Site (opposite p. 1). Download them into a PC, then use a programming deck to copy them to the PIC.

Simulating the PIC

The software supplied with a programming deck usually includes its own assembler and, as already mentioned, may have a BASIC or C compiler built in to it. On the computer screen, type in the program (or part of it) using a text editor. This is usually supplied as part of the programming software, or use *Notepad*, which comes in the *Windows* package. Next comes the **simulator**.

The simulator is another item which is provided with the programming software. It runs on the PC but behaves just like a PIC would do. It can run a program at full speed or step though it line by line.

While the simulator is running, special panels on the screen display the contents of all the registers of the simulated PIC. It is easy to see what is happening and whether or not the program is working as intended. The screen also displays the contents of the simulated PIC's memory. With the more advanced simulators you are able to set up virtual input and output devices such as LEDs, digital displays, and push buttons and see how the simulated controller interacts with these.

Now is the time to look for and correct errors. When all is finally checked, click on a button to download the assembled program into the PIC. The assembler software converts it into machine code and transfers it to the PIC's program memory. Once the program is in the PIC it stays there for years (or until you change it). Plug the PIC into its socket in the robot circuit and test it for real.

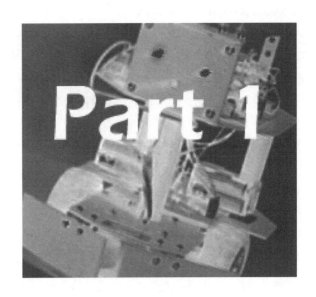

Robot Behaviour

1 Robot Behaviour

What shall it do?

This is the question we have to answer before we begin to program it as described in Part 5. What it can do depends on the type of robot it is. Mobile robots are discussed first. For gantries, turn to p. 16. From p. 17 onward we discuss the activities common to all kinds of robot.

Activities of mobile robots

Moving around

By definition, all mobile robots move from place to place. They need to be able to move forward, to reverse, and to turn to the left or right. Robots are often operated in confined spaces so it useful to be able to spin on one spot. Variable speed is less important and often unnecessary.

The *Quester* (Project 4, p. 258) runs on three wheels. Two of these, to the left and right, are the drive wheels. Each has its own electric motor. The third wheel is a castor, used for balance.

The panel on the right tells you where to look for details.

The *Scooter* (p. 165) also has three wheels, but uses only one motor. Its steering is somewhat erratic but very easy and cheap to build!

Mechanics
Wheels 39
Motors 46

Electronics
Motor speed control 93
Motor direction control 94
Servomotors 98
Stepper motors 98

Programming
Steering a mobile robot 144

The chassis of the *Android* has two motors, one for the rear drive wheels and the other for steering.

Detecting and responding to light

Sight is probably the most important of all human senses. The same applies to mobile robots. Some can detect a lamp which is several metres distant, and aim themselves towards it. Or maybe they will go in the opposite direction, to end up in the safety of a dark corner.

The *Quester* robot homes on a source of light.

The important feature of light is that it is detectable at a distance. This makes it ideal for long-range sensing.

One of the problems with using light sensors is that they may be confused by room lighting or sunlight. Pulsed light sources are one way out of this problem.

The panel on the right lists where to look for descriptions of a range of light sensors. These references explain how to build the sensors and how to program the robot to make use of them.

Electronics

Programming

Proximity

Another use for light is for proximity sensing. Proximity sensors tell the robot when it is near to, but not actually touching an object. The word 'object' includes immovable objects such as walls and furniture. In proximity sensing, the source of light is not separate from the robot, but is mounted on it, often aimed in the forward direction.

Electronics

Programming

A light sensor detects a nearby object by detecting the light reflected back from it. If the intensity of the reflected light exceeds a certain level, the robot knows that something is there. A light detector aimed sideways can be used to keep a robot at a fixed distance from a wall.

Wall-following is a common type of behaviour; it is often used by maze-solving robots.

Ultra-sound is an alternative to light in proximity detection. This requires a more complicated circuit but is not subject to interference from extraneous light sources.

Ultra-sonic sensors can be programmed to measure distances, which makes it possible for the robot to map its surroundings and more easily find its way about. However, although such applications are very interesting to attempt, they are not infallible!

Contact

By this we mean physical contact between the robot and an obstacle such as a fairly massive object or a wall.

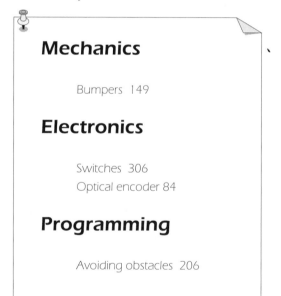

Mechanics

Bumpers 149

Electronics

Switches 306
Optical encoder 84

Programming

Avoiding obstacles 206

Typically, the robot has bumpers or possibly wiry 'antennae' arranged so that they are touched when the robot runs into anything. The usual response is to reverse a short distance, turn slightly to left or right, then move forward to try again. If the robot has a pair of bumpers, at front left and right, it is possible for the robot to work out which is the best direction to turn.

Side-mounted bumpers can be used for wall-following, instead of a proximity detector.

Other uses for contact detection occur when a robot is designed to sweep an area clear of light objects, or to find and pick up objects.

The fact that a robot is in contact with a sizeable object can often be inferred by monitoring its motion. If the drive motors are swtched on, but the drive wheels are not turning, it is likely that the robot is pushing against an immovable obstacle. A tachometer is used to determine if the wheels are turning or not.

The robot detected the box when its right bumper hit it.

Line following is a special form of contact behaviour. The robot stays in contact with a line painted on the surface over which it is moving. Line following requires two simple light sensors and the programming is easy. It is one of the most reliable techniques for guiding a robot from one place to another.

Communication

Most robots need to interact with humans, and those programmed to play games interact more than most. The robot sends messages to the human by flashing LEDs or bleeping.

Electronics

Sound sensors 85
Radio 100

Programming

Human-Robot comms
in games 232-234

Communication in the opposite direction is usually a matter of pressing a button or closing a switch.

Another technique uses a sensor that is triggered by sound.

Radio is a way of communicating with another robot to exchange information and coordinate their activities.

Navigation

Given that a robot is mobile, it seems reasonable for it to know where it is. In practice, this is not as simple as it sounds. There are basic methods of navigation, such as line following, wall following, and homing on a light source. These need the fewest sensors and are the simplest to program. They are are fine for most purposes.

Some operations require the robot to move around in 'free space', without reference to lines, walls or beacons. It might be thought that switching on the drive motors for precisely controlled periods would give good positional information. This does not work in practice. For one thing, the two drive motors do not run at *exactly* the same speed, even if they are of the same type. With both motors running forward, the robot moves forward but veers slightly to the left or right. When turning, it is not possible to control the turning angle precisely. Errors of this kind are cumulative and it is not long before the robot completely loses its bearings.

We can counteract the differences between motors in several ways. One way is to use a tachometer to count the revolutions and part-revolutions of each drive wheel. Another way is to use stepper motors instead of ordinary DC motors. However, even these may not entirely solve the problem. Depending on the nature of the surface and of the tyres, there is inevitably a small amount of slipping. This occurs most when starting, stopping or turning, and is cumulative. There is little to be done about this.

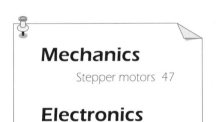

Mechanics

    Stepper motors 47

Electronics

    Stepper motors 98
    Tachometer 87

Programming

    Magnetic markers
    338

The best solution is for the robot repeatedly to take its bearings. The advantage is that errors are not cumulative. This disadvantages are that three light beacons are needed, which make setting up dfficult. Also, the maths is complicated.

One novel solution is to use a magnetic compass. Inexpensive compasses with electronic output are available from some robotics suppliers. Unfortunately, their precision is low but they are fun to experiment with.

The most practical solution is a gantry, described in the next section.

Activities of gantry robots

A gantry robot operates over a clearly defined rectangular area. It picks up objects from any point in the area and sets them down at another point in the area.

The tool (often a gripper) is suspended from a small trolley-like frame, and can be lowered and raised. The frame has wheels and runs on a pair of rails so that it can travel from one side of the area to the opposite side. This set of rails is on a larger frame at right angles to the first set, so the smaller can be moved to any point within the area. Thus the location of the tool is defined by two coordinates, its x-position and its y-position.

It is easy to design sensors that can read the x and y coordinates and a gantry robot is therefore much easier to program for applications that require precise navigation.

The *Gantry* robot solves a maze, using its laser pointer to follow the path.

Gantry robots are used in industry when very heavy loads are to be handled. The hobby versions are suited for less strenuous tasks. They are excellent for playing board games such as chess, draughts and checkers.

Like mobile robots, gantries can be programmed to solve mazes. But mobile robots are apt to lose their bearings. Because the travelling frames can be precisely positioned by keeping track of their x and y coordinates, a gantry robot can never lose its bearings.

When solving a maze, the *Gantry* does not run along passageways as a mobile robot does. It operates from above the maze, which is figured on paper or card. A narrow laser beam is projected down from the frame to mark its location.

Feedback

As an example of feedback, take an ordinary domestic refrigerator. When the temperature inside it rises above a given level the refrigerator pump is turned on automatically. It stays on until the temperature has fallen to a given level. In this way the temperature inside the refrigerator is held within close limits.

This type of feedback is called **negative feedback** because an *increase* in temperature results in the interior being *cooled* down. A robotic example is the op amp motor speed regulator circuit on p. 93. The speed regulator circuit depends on the electronic hardware to provide and respond to the feedback.

Feedback can also be effected by software. Imagine a mobile robot running along with a wall on its left. It has an infrared LED directed sideways at the wall and an IR sensor that receives the reflected radiation. The programmed behaviour is designed so as to keep the amount of reflected IR constant. In this way it keeps the robot at a constant distance from the wall.

When the robot veers toward the wall, the amount of reflected IR increases. The sensor detects this increase. The program responds to this increase by steering the robot to the right, making it veer away from the wall. The reverse happens as the robot veers away to the right.

The usual function of negative feedback is to hold things *constant*. It produces *stability*. There is also **positive feedback**. In the case of the wall-following robot described above, suppose that by mistake the output lines to the motors were swapped. Then the slightest deviation from the correct robot–wall distance would cause dramatic results. The robot would either veer permanently away from the wall or crash into it. Positive feedback results in *instability*, which is something to be avoided in robot behaviour.

There is another type of feedback that is virtually essential in a robot system. For example, a bulldozer-like mobile robot has a pusher in front of it for playing 'football'. This is normally lifted high above ground but is lowered almost to ground level when the robot sees a ball ahead of it. It must be near to but not actually touching the ground, for it might catch on irregularities in the surface.

To solve this problem, the pusher is raised or lowered by a motor which winds or unwinds a length of cord wrapped around its spindle. The switches are microswitches, the kind of switch most suitable for this kind of mechanism. If the motor is made to wind in the cord (turning clockwise in the diagram), the pusher is raised until its supporting lever touches against the lever of switch 1. This closes the switch and a signal is sent to the controller telling it to turn off the motor. If we did not have this system in place, the motor might continue turning until the pusher was damaged or the cord snapped.

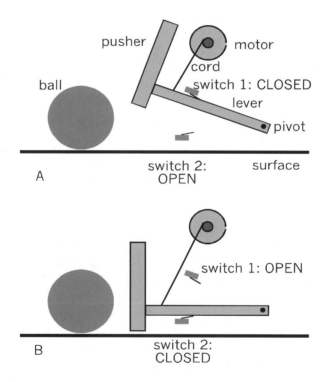

A mechanism for raising and lowering a pusher attached to the front of a robot (the robot and supporting structures are not shown).

The mechanism has a second switch which detects when the pusher has been lowered to a position just a little way above ground level. Switches used in this way, to detect when part of a mechanism has got as far as it can be allowed to go, are called **limit switches**.

In a similar way, path-following is a matter of moving the robot forward while checking the path sensors at very frequent intervals. Even on a straight path there may be irregularities that divert the robot from its intended course. If it has strayed, negative feedback is applied until it is on track again.

Monitoring output

Feedback from limit switches exemplifies one of the key principles of programming a robot :

> **Tell it what to do, then quickly check that it has done it.**

In the pusher example, tell the robot to raise the pusher, then program a loop to check switch 1 repeatedly until it is raised. When the pusher is raised far enough, switch 1 closes. The input to the controller changes and it then stops the motor.

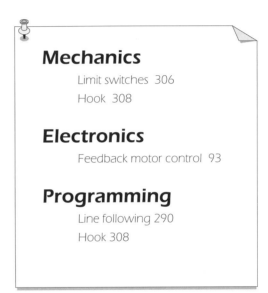

Mechanics
Limit switches 306
Hook 308

Electronics
Feedback motor control 93

Programming
Line following 290
Hook 308

Random activity

Randomness may sound an unlikely topic for a robotics book but it has its applications. A robot that is reliably built and programmed performs its tasks in an orderly and inflexible manner.

Humans are not normally like this. Indeed, if people are too rigid in their behaviour we may complain that they are 'acting like a robot' or that the person is 'an automaton'.

If a robot is programmed to run for a short, randomly-chosen distance, then turn through a random angle, and continue indefinitely in the same routine its path is totally random. We say that it is literally performing a **Monte Carlo Walk**. We say 'literally' because the same term can be applied to other random sequences that are not actually walks.

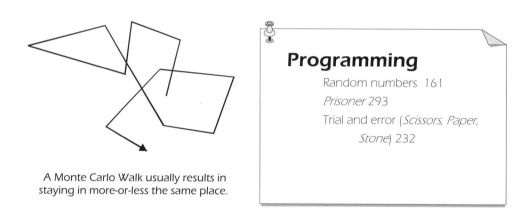

A Monte Carlo Walk usually results in staying in more-or-less the same place.

Programming

Random numbers 161
Prisoner 293
Trial and error (*Scissors, Paper, Stone*) 232

Usually we do not aim for total randomness. For example, a robot detects an obstacle in its path, backs up, turns slightly to avoid the obstacle and then continues forward. The stopping, backing and turning are fixed responses. Whether it turns left or right on a given occasion is determined as random. We can not predict which way it will turn next. This intoduces randomness into its behaviour, but not too much.

Random behaviour is produced in the software, using a random number generation routine. Actually its output is not genuinely random, but a predictable series of values that repeat after such a long interval that it appears to be random. This is actually **pseudo-random**.

Randomness has other and more serious applications. A robot that is solving a maze may be programmed to make a random choice whenever it has to go either left or right at a junction. If it is also programmed to remember which choices it made at each junction and which choices took it successfully to its goal, it can eventually learn to run the maze correctly.

This is the basis of learning by **trial and error**. The same type of learning technique can be applied to other learning tasks.

Subsumption

Imagine a mobile robot that is programmed to home on a source of light. It is programmed with a homing behaviour. The drawing below shows its path. But when it reaches point A, its proximity detector detects an obstacle between it and the source of light.

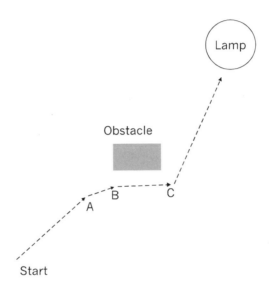

An example of subsumption.

It immediately stops its homing behaviour. Instead, it enters a phase of avoidance behaviour. The homing behaviour is **subsumed** by the avoidance behaviour.

It turns and proceeds to B. There, it still detects the obstacle, so turns more, heading for C. It has avoided the obstacle and, once it is at C there is a clear path to the lamp. It resumes its homing behaviour.

Subsumption of one behaviour by another is used as a programming technique in all types of robot. After all, it is a very human characteristic. If the phone goes while we are eating lunch, we stop eating, answer the phone and, when the call is ended, resume the meal.

Input and output requirements

This is nearly the end of Part 1, and probably the designer has developed an impressive specification. Usually this means a host of sensors and actuators, each requiring one or more connections to the controller. Now is the time to take stock, to check on the feasibility of the specification. Possibly the PIC (the controller) that was intended for the project does not have enough input and output channels.

As an example, take the PIC16F84, one of the most commonly-used of the PIC family. It has 13 input/output (or I/O) channels. Are these enough? Thirteen pins sounds quite a lot, but tank type (two-motor) steering takes four channels as outputs to control the motors. Each LED on the robot requires another channel as output. Sounding a siren takes another channel. And perhaps it is to operate a gripper. At least one channel is required for this, bringing the total number of output channels to seven. Only six left for sensors!

The simplest sensors, such as bumpers, require only a single input, but a robot generally has more than one bumper. A basic light sensor needs one channel, either for high/low digital input or for an analogue input. A proximity sensor may need one channel to signal that there is an object ahead and another pin (an output) to reset the sensor. You have already run out of pins!

The box on p. 24 leads to some solutions. One of these is to use a controller with more I/O channels. This is one reason why the PIC16F690 was chosen for this book. It has 20 pins and all except two can be used for I/O. Another solution is explained in the next section.

Distributed processing

Humans can do more than one thing at a time. For example, a programmer's heart beats at a regular controlled rate, at the same time as the programmers's fingers are pounding on the keyboard. And they may be typing with one hand while drinking a cup of coffee with the other.

OWEN'S FAILURE TO INSTALL AN OFF BUTTON WAS STARTING TO LOOK LIKE A MISTAKE.

Programming simultaneous activities on a single controller is possible but difficult. One solution is split the tasks between two or even more controllers. Each runs independently, except for occasional **handshaking** signals sent from one to the other to tell it what it is doing. This is known as **distributed processing**.

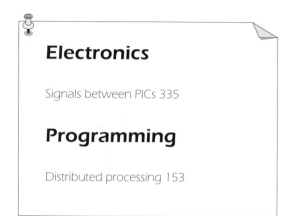

Electronics

Programming

An example of distributed processing is the *Horseshoe* game running on the *Gantry*. The controller on the main frame controls the motors and interacts with the operator. The other processor is on the x-frame. It deals with the camera sensor when scanning the playing board to register the positions of the pieces. The logic of the game is performed by the controller on the main frame.

The *Gantry* robot (Project 6.5, p. 297) uses two controllers for its more complicated activities. PIC1 is on the main frame. PIC2 is on the x-frame.

Another advantage is that two controllers have more I/O channels than one. In the case of the *Gantry*, having two controllers reduces the number of wires running between the main frame and the x-frame.

Robot Mechanics

2 Robot Mechanics

Materials

The materials for building the body or framework of the robot must be strong enough for the job, easy to work, durable and low cost. Also it should look good — have a shiny or attractively coloured surface.

Some kinds of plastic food container have all of these qualities. Project 6.1 illustrates how to build the robotic mechanisms and circuits into a ready-made box. If there happens to be a spare unused box in the kitchen cupboard, it costs nothing. The main snag is that it may not be exactly the right shape or size.

Converting a sandwich box into a robot is a short-cut way of getting into robotics, and the programs it runs can be really high-level, but a purpose-built body is more professional. The following sections describe some of the most popular materials.

Aluminium stock

Most DIY stores hold a range of aluminium stock, and it is inexpensive. It is usually sold in lengths of two metres and there is a variety of sizes and cross-sections. The drawing shows some of them.

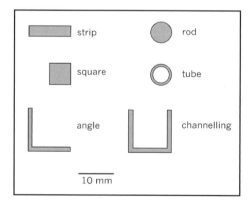

Aluminium is also available in sheets, commonly 1.5 mm thick.

This material is easy to drill and to cut, using a hacksaw. Strip and square-sectioned stock can be bent by hand, provided it is not too thick. So can rod.

It seems too obvious to point out that aluminium has the advantage of being a low-density metal. Lightweight yet rigid frames mean that low power motors can be used to move them. This in turn means that low-power batteries are needed to drive them.

Project 6.3, the *Gantry*, is an example of an aluminium framework. This was built using only two kinds of stock, strip and channelling.

Brass stock

This is useful for some of the smaller parts of mechanisms. It is obtainable from model-making stores. Brass is available in most of the same sections as aluminium stock, but in smaller dimensions. It is often sold in 200 mm lengths. Brass is more expensive than aluminium but fortunately we do not need a lot of it.

Brass is easily worked with drill and hacksaw. The thinner stocks can be bent by hand. The photo of the gripper on p. 311 shows how it can be bent to form jaws.

Its distinguishing feature is that, it is reasonably rigid but has a degree of springyness that aluminium does not have. This is why it or similar alloys are used for electrical contacts, and various kinds of spring clip.

Plastic

Model shops stock a wide range of plastic rod, tubing, angle, channelling and sheets. These are are in small sizes, being intended for scale models, but can be useful. Usually they are high impact polystyrene and special adhesive is used when building up boxes and frames.

Another source of plastic parts is the DIY store. The plumbing department stocks a range of tubing and other plumbing parts that can be used in robot building.

Examples of plumbing parts are the plastic pipe caps used as light shields for the IR sensors of the *Quester* (p. 273).

The gardening department of the DIY store may provide handy plastic tubing used in garden reticulation systems. The spacers that separate the decks of the *Quester* are cut from long PVC riser tubing.

One of the more generally useful materials for robot construction is 3 mm expanded PVC board. It is often used by signwriters and a visit to a local signwriter may provide some offcuts. If this fails, try a local plastics company. The board comes in sheets about 1 m by 2 m.

Unlike expanded polystyrene, which is soft and crumbly, expanded PVC is firm. Yet it has a certain amount of compressibility which means that nuts and bolt-heads sink a fraction of a millimetre into the surface when tightened. This makes them less likely to be loosened by vibration. The sheet is easy to drill and cut. A steel rule and sharp craft knife are all that is needed for cutting straight-edged pieces.

The sheet is manufactured in a range of attractive colours, The *Quester*, for example, is bright tomato-red.

Foam board is a similar material. It consists of a 5 mm thick sheet of solid plastic foam coated on both sides with a plastic film. It is white on one side and coloured on the other. The board is not quite as strong as expanded PVC, but is just as easily worked and suitable for small lightweight robots, such as the *Scooter* and *Android*. Robot bodywork and other structures can be assembled by using craft glue, as explained on pp. 213-215. There is plenty of scope for givng the robot a really unique appearance.

Foam board is sold at office materials stores. The brand (Elmers) used for the *Android* is supplied in sheets that measure 568 mm × 762 mm, which is a convenient size.

Wood

Wood is rarely thought of as a robot-building material but, at times, it can be just what we need. It is strong for its weight and easily cut, drilled, painted, carved and glued.

As well as a wide range of building and joinery timbers, mostly too large for robot making, DIY and model-maker's stores sell a very easily workable timber known as Balsa. Because of its low density and easy workability it is a firm favourite with flying model aeroplane constructors. For the same reasons it is useful for robot building too. The *Android* shows one way of using it.

Fixings

These hold the parts of the robot together — mostly nuts and bolts.

Buy in a stock of **nuts and bolts** in the sizes most suitable for small structures. The most generally useful size of bolts is M3 (3 mm diameter) and you need nuts to fit. Occasionally a smaller size is required. For instance, a motor may have mounting holes with M2.5 or M2 threads. Small parts such as microswitches may have 2 mm unthreaded mounting holes.

You need an assortment of the different lengths. The 10 mm and 15 mm sizes cover most needs, but sometimes longer bolts such as 25 mm are wanted, and a few of the 6 mm size.

Washers have several different functions. Plain washers, placed next to the head of the bolt, help to spread the load at that point. They are useful when bolting a relatively massive item, such as a motor, to a relatively flexible panel. Shake-proof and spring washers help prevent the nuts from loosening. Use them for bolting metal parts to other metal parts. They are not needed when bolting to expanded PVC sheet and the material itself is suitably springy.

Nylon nuts and bolts, from electronic parts suppliers, are necessary if there is a risk of the bolts causing a short-circuit. This could happen if a circuit-board is bolted to a metal panel. In such cases use nylon bolts and nuts or plastic stand-offs.

Spacers are short tubes, length 6 mm to 38 mm, made of metal or nylon. They are intended for holding a circuit board clear of the panel on which it is mounted but have several other uses. We sometimes refer to the shorter ones as collars.

There are dozens of kinds of **adhesive**, of which we employ just three. For routine fixing, general adhesives such as *UHU®*, *Bostick®*, or similar products are our standby.

Another general glue, which sticks expanded PVC sheet and Foam Board is a variety of craft glue called *Sticky Craft Glue*, made by CraftSmart. It is milky when applied but dries clear. Clamp the pieces under slight pressure while the glue sets. Examine it from time to time at first to check that the pieces have not slipped.

Super glues are quick setting and strong. We use a variety of this known as *Fix-Lock* anaerobic adhesive. A drop applied to a nut and bolt runs into the narrow space between them and sets hard. This locks the nut on to the bolt, preventing it from working loose. A locking adhesive such as this is invaluable when building robots from metals parts. Although it holds the nut secure, a little force with a spanner will loosen it if necessary.

It should really be classed as a tool but it seems more sensible to describe it along with the adhesives. The tool is the **glue gun**, which melts glue sticks and has a nozzle for applying the molten glue to the workpiece. A glue gun is a handy tool to have on the workbench for all kinds of gluing jobs.

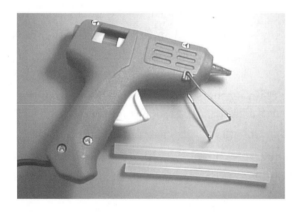

The glue sticks are melted by the electric heating coil in the gun. Press the trigger to extrude molten glue from the nozzle.

Velcro would seem to have little to do with robots but in fact it can be very good at fixing things that can not be fixed by nuts, bolts or adhesives. *Velcro Sticky Back* tape consists of the usual 'hook' and 'eye' tapes with strongly self-adhesive backs. Typical AAA and AA battery holders have no mounting holes, and there is nowhere they can be drilled to take a bolt. We use this tape for fixing battery holders and similar items.

Last but by no means least, be sure to have a pack of **Blu-Tack** to hand, as well as a packet or roll of double-sided self-adhesive tape.

Tools

The tools you need for constructing robots partly depend on the materials you use. For Foam Board the main tools are a steel ruler, a craft knife and a plastic chopping board (use one discarded from the kitchen) or cutting mat. You need a few other tools for mounting the motor and circuit boards. For building an aluminium framed robot such as the *Gantry*, a drill press is almost essential and so is a hacksaw. When you have decided what materials are to be used, select your tools from those described below.

Cutting tools

A **junior hacksaw**, with a 150 mm long blade is good enough for most jobs, such as cutting wood or plastic, and for circuit boards. For cutting aluminium or brass stock a **regular hacksaw** is faster and gives a straighter cut. If you have problems with cutting things square or if you need to cut at a particular angle, a **mitre saw** is a great help. It keeps the saw blade vertical and perpendicular to the length of the workpiece. It has gauges to help cut pieces to equal lengths. The frame that carries the blade can be rotated to cut at angles other than 90°, the angle being settable on a graduated scale. A mitre saw is almost essential for building the *Gantry*.

A mitre saw helps keep everything 'square'.

Use a medium-sized fine **flat file** for smoothing off cut edges. A set of **needle files** is useful for enlarging holes and shaping small parts. Use a **file saw** for shaping larger holes, and many other tasks. The blade is a coarse round file about 3 mm diameter and 175 mm long. It is mounted in a handle. The file saw cuts quickly and is suitable for cutting metal, wood or plastic.

A file-saw, a reamer and a junior hacksaw.

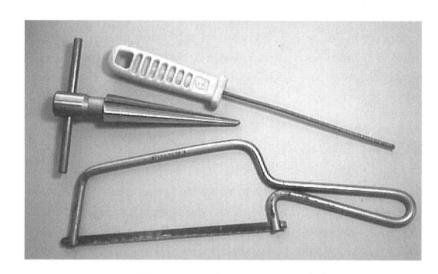

A **reamer** can enlarge circular holes up to 18 mm in diameter. It is not an essential tool, but does the job neatly. While on the subject of cutting large holes, consider getting a circular **hole-saw** that attaches to an electric drill. It is supplied with a range of interchangeable blades in diameters from 25 mm to 53 mm. Again, this is not an essential tool, but is the best way of cutting large holes quickly.

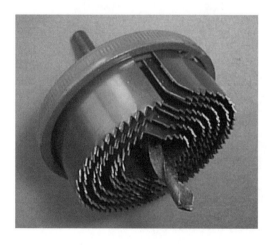

The saw blades cut holes of a range of sizes, centred on the hole first drilled by the bit.

Drills

Although a hand-turned drill is adequate in many ways, an electric drill is a boon when there is much drilling to be done. Robot building seems to require a lot of it. If you already have a small power drill for jobs about the house, it may not be worth while to get anything more professional. A drill press is not expensive, is so much easier to use and produces better results.

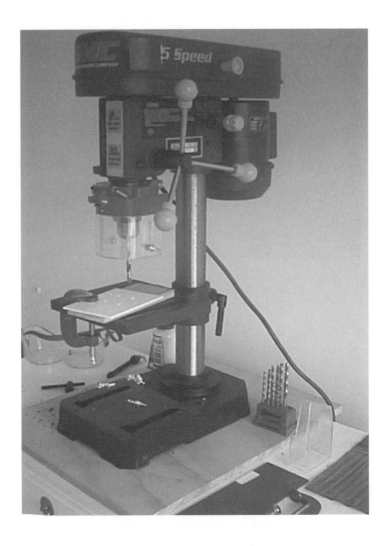

A drill press helps put the holes in the right places and at right angles. Use it for aluminium, brass, wood or plastic, preferably running it at its lowest speed.

The drill press in the photo has a drive belt transmission for varying the speed. For drilling plastic it is best to run it at its slowest speed.

There is a safety guard to prevent flying pieces of metal from damaging your eyes, but we have yet to see any flying pieces. However, wearing a pair of safety goggles is a sensible precaution, especially when drilling into metal. A pair of heavy-duty gloves ais another safety measure.

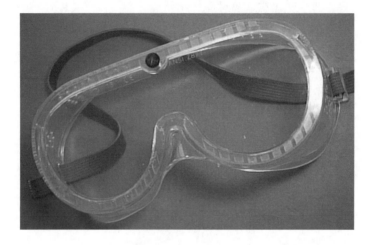

Plastic safety goggles protect the eyes when drilling.

A selection of small clamps is handy for holding the workpiece while drilling. You also need a set of drill bits, with diameters ranging from 1 mm to 6 mm. A centre-punch and hammer are used for punching the spot where the hole is to be drilled. If the workpiece is not punched beforehand there is a tendency for the bit to skid away and start the hole in the wrong place.

A small hobby drill, driven by a low voltage motor is needed for the finer work. It can be powered from the low-voltage suppy of the electronics bench. Most work on a wide range of DC voltages from about 6 V to 15 V. Kits of tools for these drills include drill bits, mini reamers, wire brushes, and felt polishing discs. It may even have a mini circular saw. The most essential items are the bits, ranging in diameter from 0.8 mm up to 2 mm, and the reamers.

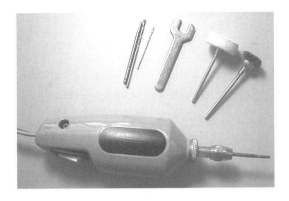

This miniature electric drill is supplied with a range of tools.

Gripping tools

In this category come a pair of standard bull nose pliers and a pair of long nose pliers. It also includes two sizes of adjustable spanners, with jaws opening up to 12 mm and 36 mm.

There are many kinds of tool for picking up and holding small items, and the most generally useful is a pair of forceps or tweezers. Get a fairly robust pair (tines about 2 mm wide) and use them for handling small nuts and bolts, bending wire, levering ICs out of sockets and dozens of other tasks. Our forceps are used more than any other tool.

On the larger scale a small bench vice holds items being sawn or drilled with the mini drill. The low-cost vices from hobby stores are made from plastic and have a suction base for temporarily attaching them to the work bench.

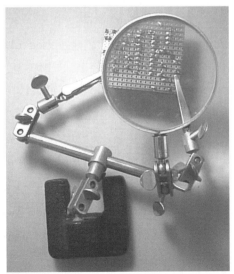

The magnifier/gripper tool shown in the photo is another handy item, particularly for holding small circuit boards and components such as plugs and switches when soldering.

Other tools

Screwdrivers are obviously essential but only small sizes with regular blades up to 4 mm wide. Metric bolts usually need a small-size Posidrive screwdriver. A set of jeweller's screwdrivers is useful on occasions.

An assortment of small files of various shapes is handy for smoothing cut edges and shaping parts of mechanisms. A set of needle files is also worth having.

An engineer's steel rule 300 mm long, graduated in millimetres, completes the tool kit.

Planning a mobile robot body

Things to think about before starting on the building.

- Chassis:
    - ◊ Ready-made chassis such as a plastic food storage box. Easy and quickly built. See Project 6.1, the *Scooter*, p. 165.
    - ◊ An existing toy. Generally easy and quick, but there could be snags. See Project 6.3, p. 246.
    - ◊ Chassis made from sheet plastic or foam board. Easy but takes longer. Gives scope for invention. See Project 6.2, the *Android*, p. 209 and Project 6.4, p. 258.
    - ◊ Chassis made from sheet aluminium. You need tools and experience to make it.
- Wheels:
    - ◊ How many? Three is a popular number.
    - ◊ Drive (traction) wheels at front or rear.
    - ◊ Steering by tank method (two independent drive wheels, two motors).
    - ◊ Steering by automobile method, one motor, with differential gear (complicated).

AFTER TWO YEARS OF HARD WORK STEVE WAS SLIGHTLY DISAPPOINTED WITH THE PUBLIC'S REACTION TO 'ROBO-BUDDY'.

- The terrain it is to run on:

  ◊ Hard, level, smooth surface such as wooden or tiled floor, or concrete or brick paving. Wheels may need composition treaded tyres for good grip.

  ◊ Carpeted floor. Mount wheels low down on the chassis to give clearance between the chassis and the carpet. Rugs are often difficult to negotiate, especially rugs with fringes. Wheels of larger diameter ride on to rugs more easily. Steps and stairs are really difficult — instead of wheels use tracks or abandon wheels and go for legs.

  ◊ Lawn. Large diameter wheels, with good clearance.

  ◊ Sloping terrain. Make the robot broad and squat, so that it does not overbalance.

- Size:

  ◊ Small size (under 200 mm in all directions) is good for robots run at home. They can find their way between the furniture more easily.

  ◊ Small size may not provide enough space for battery, motors, all the sensors and actuators that are intended.

  ◊ Small size may lead to cramped conditions, making it hard to access circuits for testing, and difficult to insert and remove the PIC microcontroller.

  ◊ Large size leads to greater weight, need for stonger chassis, need for more power to drive it, more powerful (larger and heavier) motors, more powerful (larger and heavier) battery — the situation can get out of hand! Small is beautiful!

- Shape:

  ◊ A robot that is about as long as it is wide turns much more easily in a confined space (such as the average family room).

  ◊ A robot that is much taller than it is wide is more likely to fall over on uneven surfaces, if it hits an obstacle, or when it accelerates or decelerates quickly.

  ◊ If the robot is to have jaws for picking up a load, it needs to have a broad base for stability.

There are more detailed design points in the next two sections.

Wheels

Road wheels

The main points about a road wheel are its diameter and the nature of its tread. A larger diameter is better on a rough or uneven surface because the wheel can more easily ride up over ridges and is less likely to get stuck in grooves. Also it allows there to be a larger clearance between the surface and the underside of the chassis.

If the surface is smooth and even, for example the rails of a gantry, small wheels have the advantage of light weight. It is all too easy for a robot design to finish up by being heavier than the motor can drive. Using small wheels helps to avoid this.

Tyres help the robot to run without slipping. The simply programmed robots usually start and stop abruptly. This leads to skidding or slipping. We rely on a robot being able to run in a straight line but slipping makes it run in irregular curves. Unless it is continually taking its bearings from a fixed landmark, it soon gets lost. Wheels slip, even when they have tyres, but tyres help to avoid serious slipping.

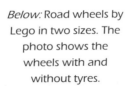

Below: Road wheels by Lego in two sizes. The photo shows the wheels with and without tyres.

Above: Road wheel and Sports tyre by Tamiya, seen on the Scooter.

Above: Road wheels and tyres by Meccano.

A recurring problem with wheels is that the hubs of the selected wheels do not fit on to the output shaft of the selected motor. There are no standard diameters. Apart from trying for a different motor or wheels, the solution is to compromise and improvise!

A wheel must be secured to the shaft so that it does not work loose and drop off, and it does not slip when torque is applied. Often a friction grip between hub and shaft is adequate, especially for a lightweight robot. If the diameter of the shaft is less than that of the hub, slip a short length of plastic sleeving, aquarium aerator tubing, or PVC insulation (stripped from a cable) on to the end of the shaft. Then push the wheel on to that. Possibly a second layer may be needed to make a tight fit.

If the wheel fits fairly well (so as not to wobble when rotating) slipping can be prevented by wiring the hub to the shaft (p. 170).

There are several sources of road wheels suitable for robots. Tamiya make a variety of wheels with tyres, including truck tyres and sports tyres. They also make tank tracks, which are sold complete with the wheels to run in the tracks. The wheels and tracks are boxed as kits and sold by hobby shops.

The constructional sets manufactured by Meccano and Lego include many different types of road wheel. It is possible to obtain most of these as separate parts from a model shop, or on the Web (pp. 51-52).

Gear wheels

Gear wheels are often needed for drive transmission and for moving arms and grippers. They are available as packeted kits of plastic gear wheels of a range of diameters from various manufacturers. Tamiya produce sets of gears, including motors, that can be assembled into gearboxes of many different ratios. Meccano and Lego produce gear wheels too, and the kinds of mechanism that can be built from them are shown in the photos overleaf.

Gear wheels transmit turning force by engaging their teeth. The number of teeth on the wheels matters more than their diameters. When talking about gear wheels we speak of '24t' wheels and '36t' wheels, meaning wheels with 24 and 36 teeth.

When one gear meshes with another, the speed (or rate of rotation, or angular velocity) of one gear relative to the other depends only on the numbers of teeth on each wheel. For example, wheel A, with 10 teeth, is driven by a motor and engages with wheel B, which has 40 teeth. As A rotates one revolution its 10 teeth mesh with 10 teeth of wheel B. But B has 40 teeth, so has turned only a quarter of a revolution. Meshing a wheel with few teeth against one with more teeth gives a reduction in speed.

In the example, the gear ratio is given by the numbers of teeth on the two wheels. The ratio is 40 to 10, or 4 to 1, generally written as 4:1. An electric motor turns at high speed, generally at several thousand revolutions per minute. If a motor turns at, say, 16000 r.p.m. and drives wheel A, then wheel B will turn at a quarter of that speed, which is 4000 r.p.m. As an equation:

$$\text{speed of B} = \text{speed of A} \times \frac{\text{teeth on A}}{\text{teeth on B}}$$

4000 r.p.m. is too fast for turning a robot's road wheels. The speed would be less if A had fewer teeth and B had more, but there is a limit on how small we can make A and how large we can make B. Instead we put a third wheel C on the same shaft as B so that they turn together. C has few teeth, say 10. Wheel C is meshed with a fourth gear D, with more teeth, say 40.

The speed ratios are like this:

A to B	4 to 1
B to C	1 to 1 (on same shaft)
C to D	4 to 1

The overall ratio of A's speed to D's speed is $4 \times 4 = 16$ to 1. If the motor and A turn at 16000 r.p.m., D turns at 16000/16 = 1000 r.p.m.

This arrangement of gear wheels is called a **gear train**. The train described above needs a few more wheels to give a speed suitable for road wheels, but the principle of meshing few teeth with more teeth is the same. The simplest way to construct a train of gears is to build a gearbox from two panels of aluminium sheet or expanded PVC. The panels are bolted together, usually at their corners.

Some practical examples of ways of using gears and gear trains are shown on p. 42

The small wheel has ten teeth and the larger wheel has 57. The ratio is 5.7:1, so the larger wheel rotated at about one sixth of the speed of the smaller one. Torque is increased about 6 times.

A worm gear. One rotation of the worm turns the the 58-toothed wheel by one tooth. The ratio is 57:1. Torque is high with this type of gear. Note that this is a 'one way' gearing because it is not possible to rotate the worm by turning the large gear. The large gear can be turnd only by rotating the worm. Shafts are at right angles.

A crown and pinion gear is a reduction gear with the shafts at right angles. The smaller gear, the pinion, has ten teeth. The crown has 50 teeth. The ratio is 1:5, so the smaller wheel turns five times faster than the large wheel.

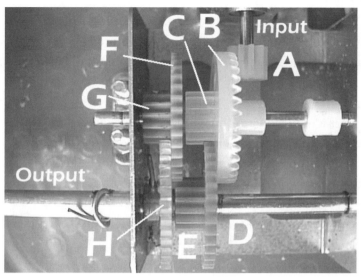

The motor turns the input shaft which has a pinion A meshing with a crown B. A small gear wheel C is moulded on to B so they turn as one. C engages with a larger wheel D, which has a pinion E moulded on it. D/E is a loose fit on the shaft. E meshes with a larger wheel F which has a pinion G moulded on it. G/H is loose. G meshes with a larger wheel H and this is a tight fit on the output shaft.
The numbers of teeth are: A = 10, B = 36, C = 14, D = 36, E = 14, F = 36, G = 14, H = 36.
The A to B ratio is 1:3.6, the other three are ratios are all 36:14 or 2.57:1. The overall ratio is 1:(3.6 × 2.57 × 2.57 × 2.57) = 1:61.

A reduction gear train reduces the speed of rotation and it increases the **torque**. This is the 'turning force', of the shaft. The amount of torque depends on the size of the applied force and its distance from the centre of rotation. Torque is measured by the sizes of the force and distance. For example, the torque of the motors used to drive the *Quester* (Project 6.4) is given as 2.1 kgf cm. It is the turning force produced by a force equivalent to 2.1 kg, acting on a wheel of 1 cm radius. A force of half that size (1.05 kgf) acting at double the distance (2 cm) has the same torque.

The torque produced by force f acting
at a distance d is
$$T = fd$$

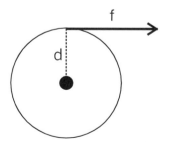

For example, a robot gripper is holding a load of 0.2 kg and the arm of the gripper is 15 cm long. The arm rotates to lift the load. The torque required is 0.2 × 15 = 0.75 kgf cm. A motor that develops a torque of only 0.5 kgf cm will simply stall when it tries to lift the load.

The secret of success is a reduction gear train, The *Quester*'s motors have built-in gearing with an 82:1 ratio. This reduces the rate of rotation of the output shaft to 1/82 times that of the motor, giving an output of 70 r.p.m. But, at the same time, it increases the torque to 82 times that of the motor (in practice, a little less because of friction).

The torque quoted for motors depends on operating voltage. The actual torque developed by a motor is less when it is run on a lower voltage.

Most applications of motors require some form of reduction gearing, but the same result can sometimes be obtained in other ways. A good example is the steering mechanism used in the robot toy (p. 249). The belt winds directly around the output shaft. The diameter of the pulley is 12.5 times that of the shaft, giving a reduction ratio of 1:12.5.

Pulley wheels

A pulley is a wheel with a grove around its rim, known as a **race**. Pulley wheels are mainly used for the transfer of force. In the *Gantry*, for example, they transfer the force of gravity to the chassis to pull it along the tracks when the winch unwinds. They are also used in the pulley system that raises and lowers the hook and some other tools. Another way in which force is transferred is by a belt drive between two pulleys.

A pair of pulleys of equal diameter simply transfer force over a distance. If their diameters differ, the result is similar to that of a gear chain. A larger pulley driven by a smaller pulley rotates at a slower speed but with increased torque.

Pulleys are connected by the drive band, but gears have to be in contact. This makes pulleys useful for transferring force over a distance. A further advantage of using pulleys is that drive bands are slightly elastic so the exact distance between the wheels is not important. With gears it is essential for the teeth of the two wheels to mesh accurately together.

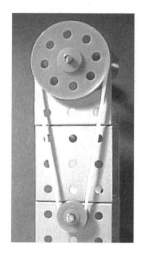

Pulleys transfer power over a distance. If the wheels are of different sizes, they turn at different speeds. The diameters of the races of these wheels are 36 mm and 12 mm.
36/12 = 3, so one turn of the bigger wheel results in 3 turns of the smaller wheel.

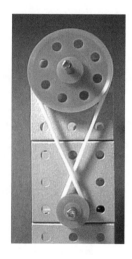

Reverse direction by twisting the drive belt. The photo shows the 'rounded triangular' shape of Meccano hubs and shafts, which prevents them from slipping. The hubs of the gear wheels are the same shape.

Shafts can be at right angles and at any other smaller angle. There is a risk of the belt coming off if the shafts are too close together.

A disadvantage of pulleys is that the drive can can come off or it may break. Another problem is that if the load is heavy the band may slip, but this is not necessarily a bad thing. It introduces **elasticity** into a system. The jaws of a gripper, for example, close on an object that is to be securely held. If they grip it loosely, it may fall out. If they grip it too tightly, they may crush it. The exact point at which the jaws should stop closing on the object is difficult to determine.

The best approach is to build a little elasticity into the system. One way is to use the gripper only for elastic objects, such as sponge balls. Another way, which allows the gripper to handle any kind of object, is to use an elastic drive band in the mechanim that closes the gripper. As it closes on the object the band begins to slip. The motor is left running to maintain the grip, so that the object does not fall out.

Motors

Although motors are electrical and therefore might be a topic for the electronics chapter, they move the mechanical parts so need to be talked about here. The discussion focuses on small low-voltage DC motors.

When selecting a motor for a project, one of the main points is its operating voltage, for this partly decides the size and weight of the battery that must be carried. Motors running on 12V need eight AA cells. These are too heavy a load for a small robot such as the *Scooter*. The *Quester* is larger and heavier so needs more powerful motors to propel it. This robot runs on a pair of 12 V motors, powered by a battery of eight AA cells. If we needed maximum power we would use dry cells, but rechargeable cells are more economical. Eight LiMH cells produce 9.6 V, and will drive the motor with sufficient power.

The next point to consider is the gearbox. It is rarely that a robot mechanism can be driven directly from a motor turning at several thousand revolutions per minute. If the motor does not have a built-in gearbox, it is almost certain that you will need to build or buy one. A built-in gearbox is neater, more convenient, and probably cheaper than a separate one.

A pair of very inexpensive motors running on 1.5 V to 4.5 V. They produce 8 gf cm and 18 gf cm respectively. These motors have no gearbox.

Finally, there is the matter of coupling the output shaft to the driven mechanism. The problem that this often raises was mentioned on p. 40. If possible, pick a motor with a shaft that matches. A 4 mm shaft is usually easier to couple. Special couplers are made for joining shafts of 4 mm and larger, but these are relatively expensive and bulky. They add to the robot's weight, which is to be avoided if possible.

A 1.5 V to 3 V motor with an attached metal gearbox. Low-cost but comparatively noisy.

A 12 V motor with built-in gearbox. Very quiet. Output shaft is 4 mm, with a flat for more secure coupling.

Stepper motors

A typical stepper motor has four sets of coils, arranged so that the rotor is turned from one position to the next as the coils are energised in a fixed sequence. This is listed in the table overleaf.

The sequence repeats and at any step, two coils are on and two are off. The sequence of pulses needed to drive the motor can be provided by a microcontroller.

Step no.	Coil 1	Coil 2	Coil 3	Coil 4
0	On	Off	On	Off
1	Off	On	On	Off
2	Off	On	Off	On
3	On	Off	Off	On

The sequence of pulses for running a stepper motor.

To get from step to step, first coils 1 and 2 change state, then 3 and 4 change state, then 1 and 2, and so on. The result of this is to produce clockwise turning of the rotor, 15° at a time. If the sequence is run in reverse, the rotor turns anticlockwise. Six times through the sequence causes the rotor to turn one complete revolution.

The rate at which it runs through the switching sequence depends on the timing of the program. The motor turns one revolution for every 24 pulses. Varying the pulse rate varies the speed of the motor. At any step, the rotor can be held in a fixed position by halting the sequence. If a stepper motor is used for driving something like an arm of a robot, the arm can be positioned exactly by programming the controller to produce the required number of pulses. There is no need for limit switches; the robot always knows where its limbs are. It can move precisely from one position to another simply by working out how many pulses to generate. Once in the required position, the rotor is, in effect, locked there and can not move. The torque required to overcome the magnetic field holding the rotor in position is several hundred gram-force centimetres.

Another advantage of the stepper motor is that its speed is precisely controllable. It is not affected by the load on the motor, except perhaps an excessive load, which might completely prevent the motor from turning. A stepper motor is less likely to stall, overheat, and possibly burn out its coils, than ordinary motor.

The stepper motor may be made to turn 7.5° per step by using a slightly different switching pattern. Motors with a 1.8° step angle are also produced. When programming a stepper motor do not feed the pulses to it too fast. At excessive speeds there is the possibility of dropped steps, which leads to an error in the positioning.

Stepper motors sound ideal for robots but we have not used them in the projects in this book. One reason is that it is difficult to program pulse production at the same time as other necessary tasks. Also the motors require large current and at least 12 V operating voltage.

Servomotors

A **servomotor** is designed to move to a given angular position. The motor has three connections to the control circuit. Two of these are the positive and 0V supply lines. The third connection carries the control signal from the control circuit, which may be a microprocessor.

A small servomotor of the kind used in flying model aircraft and robots. The 'horns' (white levers) are used for connecting the motor to the mechanisms that it drives.

The rotor of the motor has limited ability to turn. Generally it can turn 60-90° on either side of its central position.

The control signal is a series of pulses transmitted at intervals of about 18 ms, or 50 pulses per second.

The angle of turn is controlled by the pulse length:

- 1 ms: turn as far as possible to the left.

- 1.5 ms: turn to central position.
- 2 ms: turn as far as possible to the right.

Intermediate pulse lengths give intermediate positions.

Servomotors are often used in robots and in model aeroplanes and vehicles. Model shops, especially those specialising in flying model aeroplane kits, usually stock a wide range of servomotors.

Solenoids

A solenoid has a many-turned coil of wire, longer than it is wide, and a soft iron core. When a current is passed through the coil, magnetic forces pull the core into the coil. Nothing happens when the current is turned off unless there is a spring mechanism (or gravity) to move the core out of the coil again.

Solenoids are used to provide linear motion, to *pull* something in a straight line. If the core is well inside the solenoid to start with, the force on it can be quite large. A 12 V solenoid can develop a pull of one or two kilograms force. The problem is that the stroke, the distance that the core moves, is only a few millimetres. This limits the usefulness of solenoids in robots.

A miniature solenoid suitable for robotic applications.

The main applications of solenoids are short-stoke actions, such as releasing catch mechanisms. An electric door latch is operated by a solenoid. They are also built in to operate electrically controlled valves.

Construction kits

These are a rich source of ready-made parts, such as wheels, angle brackets, girders, and many more. Construction sets include Meccano™, Erector™ and Lego™. Parts are available as kits or in small quantities of individual types. Even if you intend to make the parts of your robot yourself eventually, an extensive kit of parts is handy for mocking up a mechanism during the early stages of design.

There are also the more specialist kits marketed by companies such as Tamiya. Their road wheels and pulleys are used in several of the projects in this book The local model or hobby shop is a good source of these.

Suppliers

Local stores are the first place to try, particularly:

- Model shops — for aluminium, brass, plastic and wooden stock in small dimensions. Also constructional sets and kits of parts such as wheels and tyres and items such as servomotors.

- DIY stores — for aluminium and wooden stock in the larger sizes, tools, and specialist items in departments such as plumbing and gardening, which can be put to use in unconventional ways for building robots.

- Electronics stores — for motors, solenoids, nuts and bolts in small sizes.

- Plastics retailers — for expanded PVC sheet and other materials. Look in the Yellow Pages.

New suppliers of robotic components are continually appearing and no list can be up to date. Try running a browser, such as *Google*, or *Yahoo!* Type in a general keyword, such as 'robot', 'robotics', 'Meccano', 'Erector', 'Tamiya', 'Lego', 'motors', or 'wheels'. There are lots more you can think of.

This approach hardly ever fails to produce the address of an on-line supplier or two that you had not heard of before.

Apart from advertisements in hobby magazines, the World Wide Web is the best point of contact for mail order supplies. Here are a few addresses:

www.robotstore.com — a very wide range of parts by mail order over the web.

www.legoshop.com — for Lego parts and sets by mail order over the web.

www.tamiya.com — on-line descriptions of products and a list of agents world-wide.

www.brickline.com — Lego parts.

www.robotbooks.com — many interesting titles.

www.budgetrobotics.com.

www.robohoo.com

www.hobbyengineering.com — lots of robotics parts, components, and kits.

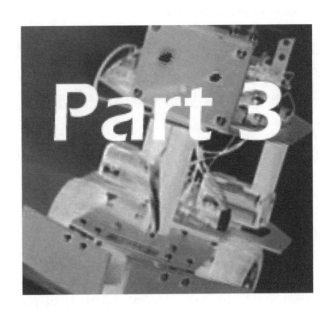

Robot Electronics

3 Robot Electronics

Materials

Circuit board

The components of an electronic circuit are nearly always assembled on a rectangle of circuit board. This is made from insulating material and has conducting copper tracks on its underside to make the connections between components. In the circuits described in this book, we use components that have wire terminals. The wires are pushed through holes in the board and soldered to the tracks on the other side of the board. Another type of component is the surface mount device (SMD) which has terminal pads that are soldered to tracks on the *same* side of the board. No holes are required. SMDs are very small and difficult to handle, so we do not use this type in the book. Take care not to ask for the SMD type by mistake when buying components.

In this book we build the circuits on **stripboard**, which has parallel copper strips, ten to the inch, perforated with holes, also ten to the inch. It is bought ready to use and allows the layout to be varied to suit the sizes of the components, and altered later if any changes are needed. It is more economical to purchase the board as a large sheet (about 300 mm long by 100 mm wide) and cut it into smaller pieces with a junior hacksaw.

The copper strip side of standard stripboard.

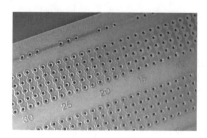

One of many more elaborate circuit boards, intended for integrated circuits (which include PICs).

Components are connected by soldering them to the strips and also by soldering connecting wires at right-angles across the board to connect some of the strips. This is known as the Manhattan layout. The drawing shows how it works.

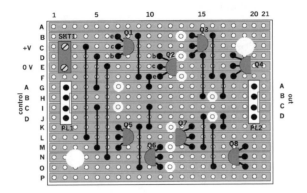

A typical stripboard layout, with strips
running east–west and connecting wires
running north–south.

While thinking about circuit board layout now is a good chance to say a few things about the layout diagrams in this book. The first point is that they are drawn as if the board is transparent. They show the components, which are on the top side, *and* the strips, which are below. They also show the circular cuts in the strips, which are not actually visible on the component side of the board. The two larger, clear circles are holes drilled in the board for mounting it on a pair of bolts.

The large circular dots are the places where leads are soldered to the strips *beneath* the board, and these too are not visible from the component side.

The components in this drawing comprise a 2-way screw terminal (top left), two 4-way header plugs (centre left and right) and four transistors.

Connecting wire

Three kinds may be needed:

- Single-stranded insulated connecting wire, 0.71 mm diameter. Used for wiring up circuit boards. Can be bought by the metre, and a metre or two in two or three different colours is worth keeping in stock.

- Multi-stranded flexible insulated connecting wire (sometimes called bell wire), 13 ×
  0.12 mm (that is, 13 strands, each 0.12 mm diameter). Use for connections to off-
  board components such as motors, and for connections between boards. Stock a
  few metres in a few different colours.

- Tinned copper wire, bare, 0.71 mm diameter. We prefer this to insulated wire for
  wiring up circuit boards. Not having to strip its ends saves time and the risk of
  short circuits is minimal. The exposed wires provide plenty of contact points for
  use when testing the circuits.

Solder

Solder is a mixture of 60% tin and 40% lead. It comes as wire with a central core of resin.
For building electronic circuits the narrower gauge, 0.71 mm diameter is preferred.

Recently introduced regulations mean that lead solder may no longer be used in certain
countries, including those in the European Union. Lead-free solder must be used instead.
This usually consists of a mixture of tin and antimony. It is available in wire form,
0.71 mm in diameter.

Insulation

It is often essential to stop currents from flowing where they should not flow. Use PVC
insulating tape to wrap around wires or terminals that are at risk. Use it also to wrap
round joints where two wires have been soldered together. A reel of red and a reel of blue
will cover most needs.

Insulated sleeving is PVC tubing a few millimetres in diameter. It is made in a range of
colours. It is handy for protecting bare wires or where two wires are soldered together.
Most sleeving is of the heatshrink variety. It fits loosely over the joint but, when heated
by holding a hot soldering iron near it, it shrinks to half its original diameter. This
anchors the insulation firmly in place.

Heatshrink tubing is sold in packs of assorted diameters and colours. A pack will last for
years.

Tools

Many of the tools listed in Part 2 are also of use for building electronic circuits. These include small screwdrivers, long nose pliers, junior hacksaw, a file or two, forceps, a mini drill, and a magnifier.

Designing tools

These are the things that are used during the design stage before you get to actually building the circuit.

A **bench power supply unit** (PSU). This should be able to supply a range of DC voltages up to 12 V regulated at a current of up to 1 A. A cheaper alternative that will cover most demands is a plug-top PSU. This is the sort that looks like a mains plug and goes directly into a mains socket.

An even more basic supply is a battery box holding the number of cells needed to produce the required voltage. This could be the battery that is intended to go inside the completed robot. Alkaline cells produce 1.5 V each when fresh. NiCD cells and NiMH cells produce 1.2 V each and are rechargeable. If you use this type you will need a battery charger. This is the most economical long-term solution for battery operation.

This plug-top PSU (with Australian plug) produces a DC output switchable to 3 V, 4.5 V, 6 V, 7.5 V, 9 V and 12 V. The maximum current is 1 A.

Circuit designs are developed and tested on a breadboard before soldering them to a circuit board. The board has an array of sockets, connected in groups. Terminal wires are bare-ended wires which are pushed into these sockets to connect them electrically.

Breadboarding is usually the first practical step when designing a circuit.

An assortment of **jumper wires** of various lengths are used with the breadboard. They are sold as sets, but you can make them yourself by cutting and stripping the ends of a few dozen pieces of *single cored* (essential) connecting wire. Let them vary in length from about 2 cm to about 10 cm.

It is a good idea to prepare a few wires stripped at one end and with a miniature crocodile clip soldered at the other. Another useful item for breadboarding and for circuit testing generally is a set of a dozen or so **test leads** in a variety of colours. These are made from light-duty flexible wire and have a hooked test clip at each end.

Another indispensable item is a **testmeter**, or **multimeter**. Preferably it should be digital and should measure DC voltages up to, say, 20 V. It may measure currents up to 1 A, but current measurements are not often made in robot circuits. It should measure resistance and capacitance. Continuity checking and diode testing facilities are well worth having.

Digital meters are favoured because of their low input impedance (they take very little current from the test circuit) and their 4-figure precision (but three figures are enough for most purposes). But a rapidly changing voltage produces an annoying and unreadable scramble of digits on the display. This is when an analogue meter comes into its own — even a cheap old model that you nearly threw away. Its wavering needle tells you almost all you want to know about voltage swings.

Soldering tools

A **soldering iron** is an essential item. A soldering station with thermostatic control is nice to have but a simple electrically heated iron will do. The main points are that it should be low power, about 15 W, and the bit should be no more than 2 mm in diameter.

If you have several models of iron to choose from, select the one with the above features and with a light-duty power lead. Some makes of iron are spoilt by having a heavy duty power lead capable of supplying a 1 kW device. The thick cable makes it difficult to wield the iron with precision.

An important accessory is a **soldering iron rest** to hold the hot iron when not in use. It should have a sponge, which is dampened for wiping the iron.

Another soldering accessory is a **heat shunt**. Designs vary but the idea is that the shunt is a block of copper or aluminium that is clipped on to the wire lead of a component when it is being soldered to the board. It is generally used when soldering semiconductor devices such as diodes and transistors.

 The shunt is clipped on between the end being soldered and the body of the component. Heat from the soldered part flows into the shunt rather than into the component.

Other tools for electronics

A **wire and cable stripper** removes the insulation from the end of the connecting wire in a single action. It saves a lot of time. But most wire strippers are designed for use by electricians. They will strip the insulation from hefty mains cable, or TV antenna cables, but not from the thin wires such as we use in electronic hobby projects. Choose with care.

Wire cutters of the **side cutter** type trim the component leads short after they have been soldered to the board or terminal. They give a neat finish to the work and no other tool does the job.

Last on our list is a **spot face cutter**. It looks like a drill bit mounted in a plastic handle and it is used for cutting the circular breaks in the copper tracks of strip board.

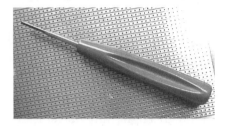

A spot face cutter.

While on the subject of cutting the strips, **ALWAYS** examine each cut with a string magnifier. It is too easy to leave a microscopically thin hair of wire around the edge of the cut. It is surprising how much current can flow through such an invisible thread. Make this the first fault to look for (with that magnifier!) when a circuit fails to work.

Components

If you are building one of the projects in this book or working to some other ready-made design, it is usual to buy just the components you need and no more. Even then it is wise to buy a few spare transistors and other items that you might damage though overheating or because of unintentional short circuits.

On the other hand, if you are designing your own circuits or modifying an existing circuit you need a small stock of the more commonly used components. Here are suggestions:

Resistors: A few of each of the E12 values from 100 Ω up to 1 MΩ. The metal film type, with 1% tolerance, and rated at 0.5 W are the best. They are usually sold in packs of 10.

E12 resistors

The values are 10, 12, 15, 18, 22, 27, 33, 39, 47, 56, 68, and 82. Next come multiples: ×10, ×100, ×1000, and so on, up to 10 MΩ. There are also sub-multiples ×0.1 and ×0.01, but these are rarely needed.

The easiest and most economical way to stock up with resistors is to buy a resistor pack that includes all the common values. Buy from a reliable source. So-called 'Bargain packs' from unknown sources may contain lots of the values that nobody wants.

- **Variable resistors:** The miniature horizontal preset type are good for breadboarding. Sometimes called trimpots. Handy values are 470 Ω, 1 kΩ, 4.7 kΩ, 10 kΩ, 100 kΩ, and 1 MΩ.
- **Capacitors:** Their main use in our circuits is for smoothing spikes and pulses out of the supply. Stock a few MKT polyester capacitors, value 100 nF.
- **Transistors:** Probably the most often needed is the BC548 for use in transistor switches, but other types can be used for this purpose. The table lists types to choose from.

Bipolar junction transistors (BJTs):

Type no.	npn or pnp	Max. current (mA)	Current gain*
BC327	pnp	800	100-600
BC337	npn	800	100-600
BC548	npn	100	110-800
BC639	npn	1000	200
BC640	pnp	1000	200
2N3904	npn	200	300

* For a collector current of 10 mA. Transistors are often graded by gain; for example the BC548C has a higher gain than BC548A ad BC548B.

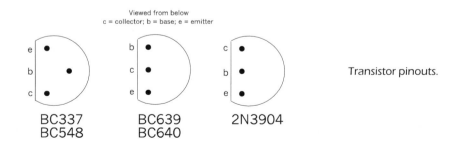

Viewed from below
c = collector; b = base; e = emitter

BC337 BC548	BC639 BC640

2N3904

Transistor pinouts.

MOSFETs (metal oxide silicon field effect transistors) are useful as switches and amplifiers when the source of the signal is unable to provide enough current to drive a BJT. As switches, they can control large currents and most types have a very low 'on' resistance. Their main disadvantage is that their threshold voltage, the gate voltage that turns the MOSFET on, varies appreciably from one transistor to another of the same type.

A MOSFET of interest for low power circuits is the VN10KM, for current up to 500 mA, an 'on' resistance of 5 and a transconductance of 0.2 S. A transconductance of 1 S means that a change in gate voltage of 1 V, results in a change of current of 1 A. The 2N7000 carries current up to 500 mA, with 'on' resistance 5Ω, and transconductance 0.1 S. Its threshold voltage is in the range of 1 V to 2.5 V, typically 2 V.

Viewed from below
d = drain; g = gate; s = source

Pinouts of two low-power MOSFETs.

- **Light emitting diodes**: A few of the standard brightness 5 mm LEDs are a help when testing the output from logic circuits.
- **Push-buttons**: Buy two or three of these, solder short single-stranded wire leads to them and strip the ends of the wires. These provide digital input to breadboarded logic circuits.

This completes the list of components that often appear on the breadboard when designing circuits for input to and output from a microcontroller. As well as the above, build up a selection of sensors and actuators such as LDRs, photodiodes, infrared emitting diodes, a solid state buzzer, and a small loudspeaker. Exactly what you include depends on your interests. Where necessary, solder short leads to these so that they are ready for used on the breadboard.

A supplier's catalogue or two, or the equivalent on a CD, will usually give technical data, will help you know what is available, and may even suggest ideas that lead to successful new designs.

Connectors

This book is based on the idea of circuit **modules**. These modules can be put together in many different ways to build a variety of robots. Modules also have the advantage that their circuit boards are small and so will fit more easily into that small space in the cramped interior of the robot. A modular system can be improved, added to and revised without having to re-build the whole system.

There are several types of connectors suitable for robot circuits. The cheapest are the 0.9 mm or 1 mm **terminal pins**. These are pushed through the holes in the strips of the circuit board and soldered in place. The end of the connecting wire is bent into a little hook, and soldered to the pin. Soldered connections are very reliable but a trouble to undo if the connections are to be changed.

PCB terminal pins are similar to the basic terminal pins but are longer and usually gold-plated. They are pushed through the holes and soldered like the plain type of pin.

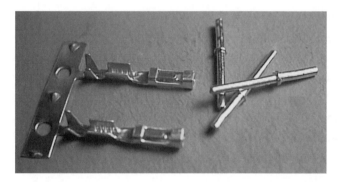

PCB terminal pins connect with push-on sockets. The sockets
are supplied in break-off strips.

The matching sockets are broken off the strip, crimped around the bare end of the connecting wire and soldered to it. Contact is good because of the gold plating and the springy socket.

The designs for several of the controller boards make use of this kind of connector. This lets the connections between the controller and its peripherals be changed as required during the development of the robot's electronic system.

Header plugs and sockets are a good way of connecting modules that have two or more connections. The plug is soldered to the board and the socket crimped and soldered to the connecting wire.

A 6-way header socket shell (top) holds the individual sockets (left) and pushes on to the header plug (right), which is soldered to the board.

The plugs are available in 2-, 4-, 6-, and 8-way versions. The plugs and sockets have polarising ridges so it is not possible to connect them the wrong way round. This makes it important to wire them the right way round as the individual sockets are very hard to extract from the shell once they have been inserted.

PCB screw terminals are another way of making firm connections that are easily changed. We find them useful for the power supply lines as it is possible to insert two or more wires in the same terminal, and daisy-chain the power lines from board to board.

A 2-way PCB screw terminal. 3-way and 4-way versions are also available. The pins are spaced 5 mm (two strips) apart.

Power supplies

The various types of power supply — batteries and PSUs — are discussed on p. 57. Here we work out what voltage the supply should have and go into more detail about how to provide it.

The first stage in planning the supply is to list the devices and circuits in the system and what they will need. All systems will include one or more PICs so start with this. The most recent PICs operate at any voltage between 2 V and 5.5 V. This wide range gives a degree of flexibility except that it does not include 6 V, which could be conveniently provided by four alkaline cells. The nearest is four NiMH (or NiCd) at 4.8 V. In practice, though rated at 1.2 V per cell they give 1.3 V when fully charged.

The system will almost certainly include one or more motors. Usually the motor is chosen for its dimensions and its running speed (and perhaps its price). Provided its operating voltage is not more than 12 V, we either run it at this voltage or on a lower voltage on which it will run fast enough.

Other devices in the system may have special requirements. For example, CMOS logic circuits of the 4000 series require between 3 V and 15 V, which is easily met, but the 74HC series need between 2 V and 6 V. They can not be run on the same supply as a 12 V motor. Some of the solid state buzzers and bleepers have a wide range of operating voltages but others have not, so check this point before you buy.

If everything can run at the same voltage it makes the circuit design much simpler. This is why it is preferable to use low-voltage (3 V or 6 V) motors that can run on the same supply as the PIC. Sometimes a 12 V motor, solenoid or relay is the only suitable type and we have to set up two supplies.

The power supply circuit, single or double, should also include an on-off switch and preferably an indicator LED to light when it is switched on. Some circuits for this are shown overleaf.

If the drive motors have their own supply it is a good idea to give them a separate switch. It is then possible to run the PIC and test it without the robot shooting all over the place.

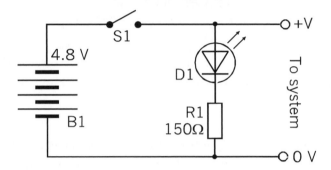

Circuit for the supply of a single-voltage system. The source is a battery of four NiMH cells. S1 is a panel-mounting toggle switch. D1 is a standard brightness LED. The resistor limits the current through the LED to about 20 mA.

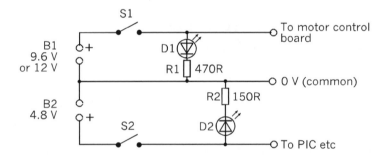

Circuit for the supply of a dual-voltage system (the *Gantry*, p. 297). There is a switch for each supply, but the 0 V rail is common to both supplies. Note the differing resistances of R1 and R2.

Calculating the resistance

There is a forward voltage drop of roughly 2 V across a conducting LED. The drop across the resistor is $(V_{supply} - 2)$. If the current through the LED is to be i amps, the resistance must be $(V_{supply} - 2)/i$.

Example: If V_{supply} is 4.8 V and the current is 0.02 A, the resistance R is:

$$R = (4.8 - 2)/0.02 = 2.8/0.02 = 140 \ \Omega$$

Use the nearest value, 150 Ω.

The controller circuit

Because the electronic system of the robots is modular, the controller board has little on it except for the PIC. The drawing shows the essentials.

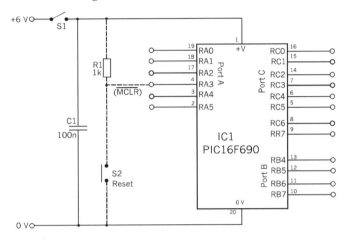

With its 20-pin package the 690 provides up to 18 I/O pins.

The switch S1 is the power supply switch seen in the drawing opposite. The polyester capacitor C1 is to absorb voltage spikes on the positive supply line.

When power is first switched on, channels RA0 to RA2, RA4, RB4, RB5, RC0 to RC3, RC6 and RC7 are all analogue inputs. The rest are digital inputs. The analogue channels can be defined individually as digital and all channels except RA3 can be defined as outputs. Outputs are always digital.

When they are configured as inputs, the channels of Ports A and B (except for RA3) can all have weak pull-ups. These can be brought into action individually. The weak pull-up acts like a high-value resistor between the input pin and the supply line. The input is read as a logic high, unless it is strongly grounded to the 0 V rail.

Channel RA3 is an exception. It can be configured as a digital input with no weak pull-up. The circuit for this is shown in dashed lines. It requires an external resistor. If the channel is configured as $\overline{MLCR}$, to reset the controller, it automatically has a weak pull-up and does not need the external resistor.

Input circuits

One-bit input

The drawings show the main types of circuit to provide digital (high or low) inputs to the controller. The simplest is just a switch between the terminal pin and the 0 V rail. Input is normally logic high, and changes to low when the button is pressed.

To use this circuit the channel must have a weak pull-up enabled. This means that it must be one of the channels of Ports A or B. The switch is drawn as a push-button, but there are many types of switch that can be used instead. These include toggle switches, microswitches (ofter used as limit switches, see p. 306), tilt switches, thermostat switches, reed switches, relays, and even devices such as pressure pads. All of these can provide direct input to a controller.

Sometimes an input channel is 'noisy', picking up interference from motors or other electromagnetic devices. Voltage spikes may overcome the weak pull-up, especially if the input line is more than about 10 cm long. In such cases, it is safer to use a stong pull-up, or perhaps a strong pull-down, in the form of a 10 kΩ resistor, as shown below.

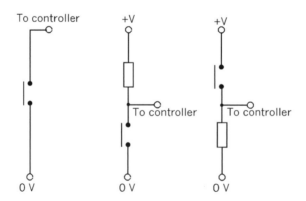

One-bit input circuits. *Left:* for use when the channel has weak pull-ups. Input normally reads as high, but goes low when the button is pressed. *Centre:* for use on channels without weak pull-ups. A 10 kΩ resistor is suitable. Input is normally high. *Right:* The same as the previous circuit except that input is normally low.

In a robot with lots of moving parts there may be many micrcoswitches acting as limit switches. They are necessary but each one occupies an I/O channel. Maybe some of these channels are wanted for other purposes. One way to be economical with channels is to logically OR some of the inputs, as shown below.

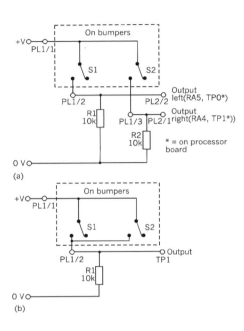

(a) Connections for two switches with separate inputs to the processor. (b) The switches are ORed together on to a single channel.

As an illustration, a robot has two bumpers, one at the front and one at the rear. They can not normally be both in contact at the same time. Use circuit (b) above to OR the switch outputs. When S1 OR S2 is closed the channel goes high. The robot can work out which was closed by knowing whether it was moving forward or backward at that time.

The logic of this assumes that there is no other robot on the scene to run into the bumpers of the first robot while it is stationary. In that case, its back to circuit (a).

There are OR versions of the input circuits shown opposite. A digital input channel can also accept logical input from a CMOS gate or other CMOS output terminal. This can often be useful if the output from a sensor does not swing fully from low to high. A gate 'squares up' the signal, producing clear logic levels that the PIC can understand.

THE COMPUTER WASN'T THE ONLY THING TO SHORT CIRCUIT.

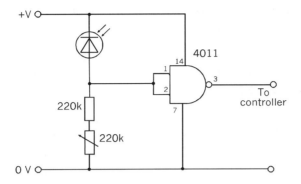

One-bit input from a photodiode (visible light or infrared).

The drawing shows a photodiode connected to a digital input channel with no weak pull-up. The diode is reverse-biased so that only a small leakage current flows through the resistors. The variable resistor is used to adjust the voltage developed across the resistors.

For a CMOS gate any input above half the supply voltage counts as logic high; an input below supply/2 counts as low. Here the gate is a 2-input NAND with its input tied together. It then acts as a NOT or INVERT gate. As light falling on the photodioide increases, the leakage current increases and the voltage at the gate inputs increases. The output changes from high to low at the half-way voltage.

This action could be effected without the gate by using the PIC's internal comparator, but there are occasions when an external logic solution, such as that given above, is preferred.

Analogue input

The analogue output from a sensor such as a photodiode (see above) or light dependent resistor circuit can be read by using the PIC's internal analogue-to-digital converter. This has a 10-bit output, so we can read the sensor with high precision. It is not just a matter of high versus low. The PIC van be programmed to do different things at different light levels.

The *Scooter* (p. 166) is an example of using an AD converter. The robot spins around, continually reading the light level ahead of it, until it locates the direction of the brightest light in its view. Then it moves forward towards it. In this robot the forward-facing LDR is connected to pin 19, which connects to the AN0 input channel. If you are planning a robot and think that you may want to read analogue input, reserve pin 19 (AN0) and possible pin 18 (AN1) for this purpose.

If we are interested in only one input voltage level, a voltage at which a given activity is triggered, we use the PIC's comparator (p. 120). With this, the analogue input from the sensor is compared with a fixed reference voltage, and the one-bit output of the comparator goes high or low depending on whether the input from the sensor is above or below the reference voltage.

The reference voltage can be provided internally, and is programmable, or it can be generated externally. The external reference may either come from a reference voltage device or from a potential divider circuit. The first of these gives a fixed voltage with high precision; the second gives a voltage that is a fixed *proportion* of the supply voltage. The potential divder is often the better one to use as it is adjustable with a screwdriver.

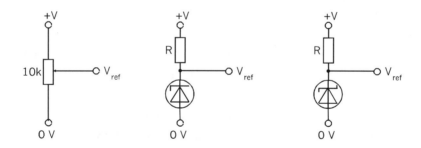

Voltage references: Left: variable output from a potentiometer is proportional to the supply voltage. Centre: Zener diode, gives stable voltage with varying supply. Right: Band-gap reference gives highest precision.

Calculating R

1) I_{max} = current required (in mA) + 5
2) Minimum power rating is $V_{ref} \times I_{max}$
3) Resistor R = $(V_{supply} - V_{ref})/I_{max}$

For the AD converter, the input voltage converts to an output of 3ffh (if all ten bits of ADRES and ADRESH are read) when it equals the reference voltage. If the supply voltage is used as a reference this gives high precision resolution, even when converting low voltage inputs.

For the comparator the voltage reference decides the input level at which the output changes state. There are two ways of setting the reference, using bits <3> to <0> of the VRCON register (p. 121). This method of setting the reference gives only 16 possible settings. When trying to set the reference to the critical level, a given step may be far too low and the next above it may be far too high. In this case the input may need amplifying before it is sent to the comparator. The diagram illustrates a way of improving the discrimination of the sensor.

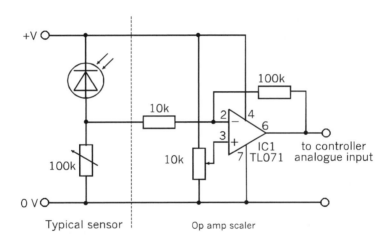

The op amp scaler circuit processes the output voltage of a sensor before sending it to the PIC.

The sensor, a reversed-biased photodiode, has a variable resistance in series with it to adjust it to operate in high or low light levels. Essentially the op amp circuit is an inverting amplifier. Its gain as determined by the two fixed resistors is $100000/10000 = 10$. A small change in the voltage from the sensor is now large compared with the voltage steps of the internal voltage reference.

The variable 10 kΩ resistor sets the zero level. Note that the output of the circuit is inverted.

Sensors

A robot needs to be aware of what is happening in the world around it. That is why all our robots are equipped with several sensors linked to the controller. This section lists sensors that are often used in robotics.

Resistive sensors respond to changes in a quantity such as light or position and their response is a change in their resistance. A change in resistance is easily measured by passing a current through the sensor and generating a changing voltage which is sent to the controller. Usually the sensing circuit is a potential divider with the sensor as one of the resistances.

EMF sensors respond to changes in a given quantity by changing the EMF (electromotive force or, loosely, a voltage) that they produce. This is sent to the controller.

There is scope for ingenuity when using sensors. Shining a beam ahead of it, the *Scooter* (for example) has a light sensor to detect light reflected back to it by an object blocking the robot's path. The light sensor is being used as a proximity sensor.

Light sensors

The most commonly used light sensor in our robots is a **light dependent resistor** which, as its name implies is a resistive sensor. The resistance of a typical LDR, such as the ORP12, ranges from 1 MΩ or more in darkness to about 80 Ω in bright sunlight. Indoors, with indirect daylight or artificial illumination, their resistance is a few kilohms.

LDRs respond to light of most colours, with a peak response in the yellow. Of all the light sensors, the LDRs are the slowest and their response times are several tens or hundreds of milliseconds. Although this is seems quite fast to humans, the PIC works much faster than this. Programs may need a short delay to allow time for the LDR to catch up with it.

The potential divider (see drawing opposite) can have a fixed resistor, a variable resistor, or both. The variable resistor allows for setting the output voltage for any given light level. The total resistance should be in the same range as the average resistance of the LDR under the expected operating conditions.

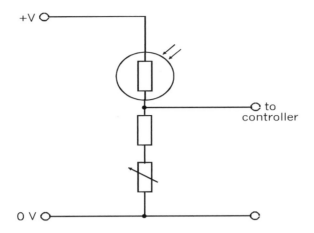

+V

to controller

0 V

As light increases the resistance of the LDR
decreases and the voltage sent to the controller
rises.

Another popular light sensor is the **photodiode**. The action of these depends on the fact
that the leakage current when the diode is reverse biased varies with light intensity. The
circuit is on p. 73. The leakage current is very small. In darkness it is only a few
nanoamps and rises to about 1 mA in bright light. The resistor has a resistance of a few
hundred thousand ohms, so the current generates a reasonable voltage across it. Often a
330 kΩ resistor provides suitable output voltage. The output must be connected to a high
impedance input so that the voltage is not pulled down. The PIC is a CMOS device so has
high-impedance inputs.

A photodiode is generally more responsive to light from the red end of the spectrum.
Some are specially sensitive to infrared. These are used with infrared LEDs for reading
optical encoders (p. 84). They are employed as sensors in line-following robots because
they are less subject to interference from external sources of visible light.

The response time of a photodiode is fast, generally a few hundred nanoseconds, so there
are no problems with this.

A **phototransistor** (overleaf) has properties similar to those of a photodiode, though their
response time is longer. They are connected in the same way as an npn transistor in a
common-emitter amplifier. Phototransistors often lack a base terminal and, if present, the
base is usually left unconnected.

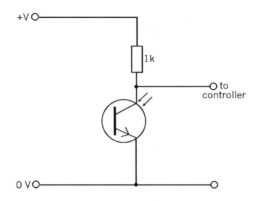

+V

1k

to
controller

0 V

Left: Some phototransistors have no base
terminal. They are 2-terminal devices with
collector and emitter.

The reason that a phototransistor can operate with a base current is that the light falling on the transistor releases a supply of electrons. These electrons act in the same way as a base current.

Phototransistors are often packaged with an amplifier circuit or a Darlington transistor output on the same chip for greater sensitivity. There are similar devices based on photodiodes.

Digital output

The analogue output from a light sensor is usually processed by the PIC's in-built comparators but sometimes there are not enough of these, and in any case it is simpler to handle the triggering level in the hardware. This circuit uses an op amp comparator to convert the analogue output into digital.

The op amp has two inputs and there is no feedback, so the difference between the input voltages is multiplied by the open loop gain of the amplifier. The output swings very sharply from low to high when the voltage at pin 3 exceeds the voltage at pin 2.

The voltage at pin 3 is always half the supply. The voltage at pin 2 varies directly with light intensity. It is set by adjusting VR1 to bring it to half supply when the LDR is receiving light of the triggering intensity.

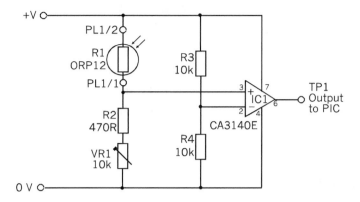

The output of this circuit goes high when the
light level rises above a preset amount.

This circuit is used for the LDR light sensor of the *Quester* robot (p. 258). The output of the
CA3148E swings close to the supply rails, providing a clear signal for the controller.

The *Quester* demonstrates another way of producing digital output. A CMOS logic gate
changes state when input voltage levels are close to half the supply voltage. They act like
a comparator with the reference voltage set at $V_{supply}/2$.

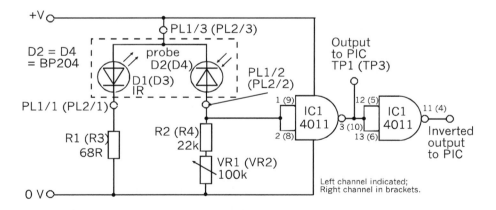

The circuit for the two IR probes of the *Quester* robot (p. 258).

The circuit uses CMOS NAND gates but could use NOR or INVERT gates instead. There are two IR probes and the component numbers of the second probe are shown in brackets.

Each probe has an IR LED (D1, D3) to illuminate the area beneath the probe. The reflected IR is detected by a pair of BP204 IR diodes. Note that the polarities of the two diodes are in opposite directions.

Detecting colours

A robot may be required to **detect the colour** of an object. For example, it may have the task of sorting building blocks of different colours. A simple way to do this is to illuminate the object with light from two or three different colours, one at a time. A single LDR light sensor measures the amount of light reflected from the object.

As a test run of the technique, small (30 mm) squares of coloured paper are illuminated by light from red, green and blue high-intensity LEDs. An LDR is exposed to the light reflected from these squares. The LDR is part of a circuit like that on p. 75, and the output voltage is sent to a comparator. With a suitable setting for its reference voltage, the comparator's output is read for each coloured paper and for each coloured LED. The output is 0 for low reflection or 1 for high reflection. Here are some typical readings:

Paper colour	White	Red	Green	Blue	Black
Red LED	1	1	1	0	1
Green LED	1	0	1	0	0
Blue LED	1	0	0	1	0

Each paper except for red and black produces a different response. If the PIC is programmed to flash each LED in turn, and to read the output from the comparator each time, the colours of the papers can be identified. Possibly the problem with black is that it is reflecting (or emitting?) infrared, to which the LDR is sensitive. Under green light, the red paper reflects less than black, so it is possible to distinguish these colours if a suitable reference voltage is used.

A light detector with memory

This circuit detects brief flashes of light. Although the PIC's interrupt facilities can respond to this sort of thing it is not necessarily convenient to have the program interrupted at an unpredictable time. This circuit allows the controller to poll its output to discover if a flash has occurred.

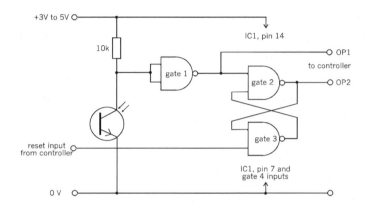

The light flash sensor is based on a phototransistor and a set-reset flip-flop.

Q1 is a phototransistor, preferably of the Darlington type. When a flash of light is received, Q1 conducts for an instant and the voltage at its collector briefly falls. This low pulse is inverted by gate 1 and sent by the OP1 terminal to the controller. If the input channel is configured to 'interrupt on change', an interrupt is generated.

Gates 2 and 3 form a bistable flip-flop. Its state is changed by *low*-going pulse at its set or reset inputs. This part of the circuit is intended to detect a break in a beam of light that is normally falling on Q1. A break in the beam generates a high pulse, which is inverted by gate 1 and triggers the flip-flop to change state from reset (output OP2 low) to set (OP high). Once made high, it stays high until the flip-flop is reset. While it is in this state the controller polls OP2 at a suitable time or times. Later the controller sends a short low pulse to the reset input, which is normally held high. OP2 returns to low.

The response of the circuit is the opposite if the IC (quadruple NAND) is replaced with a 4001 or 74HC02 IC. These have quadruple NOR gates. OP1 goes high when there is a flash of light. The flip-flop is triggered when there is a flash of light and OP2, normally high, goes low. It stays low until the flip-flop is reset by a high pulse.

Thermal sensors

You might design a robot that is attracted by the cosy warmth of your fireside, or an industrial robot might be programmed to enter danger areas and report back on the temperature there. In either case a thermal sensor is required.

The most often used sensor is a *thermistor*. This is a resistive sensor. The negative temperature coefficient type, which is the type used for temperature measuring, decreases in resistance as temperature rises. Unfortunately its response is not linear. This means that a thermistor is fine for a trigger circuit, in which a response is triggered at a preset temperature, such as a fire alarm. The sensor circuit is a potential divider similar to the one on p. 75, but with a thermistor replacing the LDR.

Thermistors are available as rods, discs or beads of the resistive material. Their resistance is usually quoted for room temperature (25°C). They are made with a range of values from 25 Ω to 1 MΩ, with a tolerance of 10% or 20%. Because their tolerance is low, a trigger circuit needs a trimmer resistor in series with the thermistor to adjust the set temperature.

Thermistor formula

The formula for the resistance R of a thermistor at temperature T is:

$$R_T = R_{ref}e^{-\beta\,(1/T - 1/T_{ref})}$$

R_{ref} is the resistance at temperature T_{ref} (temperatures in kelvin), e is the exponential constant 2.718, and β is a factor that depends on the composition of the thermistor. It is typically 4000, but the actual value is given in the supplier's data sheet.

The device for precision temperature measuring is the **bandgap sensor**. It has three terminal pins and looks like a transistor. An example is the LM35CZ, and there are several similar devices available. This device operates on any supply voltage between 4 V and 20 V.

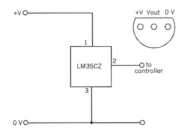

The bandgap temperature sensor needs no external
components.

The output is a voltage that is proportional to the Celsius temperature. As temperature ranges between 0°C and 110 °C, the output rises from 0 V to 1.1 V. What could be simpler?

The precision of the LM35CZ is ±0.4°C at 25°C and is no more than ±0.8°C over the whole of its range.

Motion sensors

A **tilt switch** could help prevent a disaster for a robot travelling on rough or steep terrain. The switch is mounted in the robot so that it is normally in the vertical position. The switch is open in this position. If the body of the robot tilts only a few degrees the switch closes.

The simplest way to connect a tilt switch is to a digital channel that has a weak pull-up. Input goes low when the switch is tilted.

Vibration switches are relatives of tilt switches. They make and break contact at the slightest disturbance. They are intended for use in security systems but there are robotic applications too.

A home-made tilt switch is easily constructed and may work better than a ready-made switch.

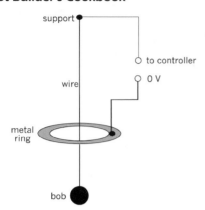

When the switch is tilted the wire of the pendulum touches the metal ring. This completes the circuit and the input channel of the controller is pulled down to 0 V.

The main problem with this is that the pendulum may continue to swing when the switch is no longer tilted. Its motion needs to be damped. Try to find a wire of springy metal or use a wire spring. Extension springs, the sort in which the adjacent turns are in contact, are better than compression springs.

Another problem with tilt switches is that they may close when the robot is accelerating or decelerating. The solution is for the software to ignore input from the tilt switch for a short time after the drive motors have been turned on or off.

A tilt switch merely indicates by its dgital output when the robot has tilted more than a fixed amount. A **tilt sensor** provides an analogue value which is a measure of *how much* it has tilted.

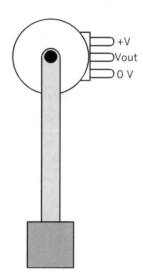

The voltage output from a tilt sensor varies with the amount of tilt.

A rigid arm is fixed to the spindle of a variable potentiometer, and the arm is directed downward. There is a mass at its lower end and this is sufficient to turn the spindle when the robot is tilted. The voltage at the wiper of the potentiometer varies with the amount of tilt and can be read with the PIC's AD converter.

This device is sensitive to tilting in a single vertical plane, perpendicular to its spindle. It may give a false or no reading when tilted in other planes.

Location sensors

These tell the robot where it, or a particular part of it, is located in space. In the *Gantry*, for example, it is essential to know exactly where the tool on the x-frame is located. The *Gantry* can not perform its tasks without this data.

For locating a robot or, more often, a part of a robot, over a range of a few tens of millimetres, we can use a technique based on a **linear potentiometer.** This is a variable resistor of the slider type, such as those often used for setting the frequency response of an audio amplifier. The ends of its track are connected to the positive supply and to 0 V. The object is mechanically attached to the slider. The voltage output at the slider varies according to its distance along the track. In short, it is a straight line servo.

There is a limit to the length of the track of the potentiometer, which restricts this type of sensor to short distances. For longer distances we can employ **markers**. An example of this technique is the way magnetic markers are used in the *Gantry*. As illustrated on p. 317, the small magnets are spaced out beside the rail along which the frame is moving. The frame carries a Hall effect sensor (p. 88) which produces a pulse when it passes a magnet. The robot knows where the frame is currently located by simply counting the pulses. It also has to know in which direction it is moving, and counts up or down accordingly.

The marker technique relies on beginning at a fixed location. This is why the *Gantry* begins by moving its frames to its base location, using the limit switches. An alternative marker technique uses optical markers, for instance regularly spaced black marks. The moving part carries a light sensor that produces a pulse as it passes a mark. The marks are counted as in the magnetic technique.

Optical encoders work on a different principle. The moving part is coupled to a transparent strip that is marked with a pattern. The pattern consists of four or more rows of bars which are either transparent or opaque. There is an array of light sources behind the strip and an array of light sensors detects light passing through the transparent bars.

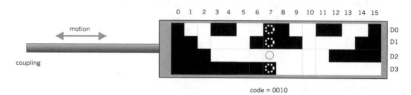

An optical encoder with sixteen positions. As shown, the encoder is at position 7 and the 4-bit code sent to the controller is 1101.

With four rows of bars and four sensors to read them, there can be 16 distinct combinations of outputs from the sensors. The pattern of bars is based on the Gray code. In this the four bits do not follow the normal binary sequence of counting from 0000 to 1111. In the binary sequence two or more digits may change at the same time (example 0111 to 1000). It is not likely that they will all change at *exactly* the same time, which causes errors. In a Gray code only one digit changes each time we move from one position to the next.

The bar encoder technique can also be used to determine angular position, using a pattern of bars in concentric circles. It may be used to read the angles between the segments of a robot arm. In a mobile robot, it may be used to read the angle turned through by the wheels as the robot moves along. Given the radius of the wheels, and assuming that the wheels do not slip, the precise distance moved can be calculated.

Limit switches are perhaps the most often used of position sensors. They are robust, reliable, and easy to read, which is why there are so many examples of them in this book.

Proximity sensors

A robot may frequently need to know if it is near to something such as a wall, the leg of a chair, or an object blocking its path. One approach is to install **bumpers** with microswitches to detect actual physical contact. The *Quester* has these.

Generally it is better to detect objects when they are near by, but before actually running into them. The **light sensors** of the *Scooter* do this by responding to light reflected back from the object. The more light reflected, the nearer the object. The amount of reflection may also depend on the size and colour of the object, so this method can give misleading results under some circumstances.

Another technique is to use **ultrasound**, as described on pp. 86-88.

Sound sensors

The circuit shown below has a microphone to detect sound, an amplifier to increase sensitivity and a trigger circuit to send a logical high output to the controller when sound is detected.

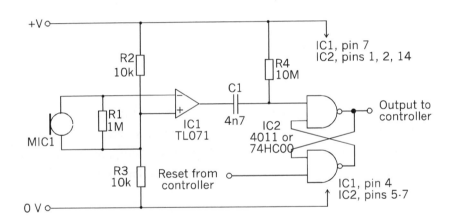

A crystal microphone generates a voltage spike that is
amplified and used to trigger a flip-flop.

The voltage spikes across R1 are fed to the op amp. This is operating without feedback, as a comparator, so its output swings widely. The amplified spike passes across C1 to the input of the flip-flop. This is based on NAND gates so is triggered by a low-going spike of more than $V_{supply}/2$. Its output swings from low to high.

Once set, even by the briefest of low spikes, the output remains high until the normally high reset input is briefly made low. This resets the flip-flop and its output goes low.

The next amplifier is more complicated to build but is much more sensitive. It is based on an electret microphone. Just a simple microphone insert is all that is needed.

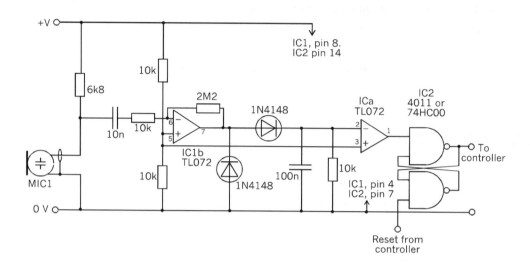

A highly sensitive sound detector with a digital output.

The audio signal generated by the microphone passes across the capacitor and is amplified 220 000 times by the op amp. The signal is then rectified by the two diodes and a positive voltage builds up on the 100 nF capacitor. This is compared with the mid-rail voltage and when this is exceeded the second op amp, wired as a comparator, swings low and triggers the flip-flop.

The build-up on the capacitor is continually discharged to the 0 V line by the resistor. Which helps to eliminate the effect of very faint background noises. After reading the output, the controller resets the flip-flop in the usual way.

Ultrasonic sensor

Ultrasound, by which we generally mean sound with a frequency of 40 kHz, has two applications in robotics. It can be used for proximity sensing and for measuring distance. There are two circuits: a generator and a receiver.

The generator circuit consists of an oscillator built from two logical NAND gates. The output from each of the gates goes to two more NAND gates each with their inputs connected together so that they act as INVERT gates. The signal at the output of one gate is exactly 180° out of phase with that from the other. These two signals are applied to the crystal of an ultrasonic transmitter and, since they are out of phase, have a push-pull action that generates a strong burst of ultrasound.

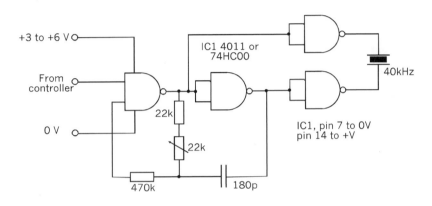

An ultrasound generator circuit,
based on a single logic IC.

The generator has a control input from the PIC, which is made high to switch the generator on. This allows the PIC to produce short pulses of ultrasound. These are reflected back from objects up to a metre or two away. The time elapsing between the production of a pulse and its arrival at the receiver is measured by the PIC. Given that sound travels at a little over 330 m/s (depending on air temperature), the time taken is used to calculate the distance of the object.

If the generator is being used only for proximity sensing and if there is a shortage of output channels, the generator can be run continuously. Omit the connection from the controller to the input of the gate. Instead, connect both of the inputs to that gate.

The circuit of the receiver (overleaf) is more complicated since it has a two-transistor amplifier to amplify the signal from the receiver crystal. The signal is then rectified by the diode to produce an output voltage level that falls when ultrasound is being received. This analogue signal is fed to a comparator or AD converter in the PIC.

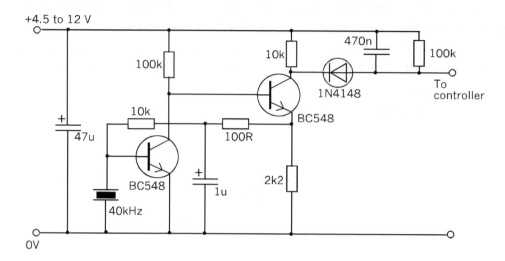

An ultrasonic receiver circuit.

Magnetic field sensor

The Hall effect device, which detects magnetic fields, has many applications. Its output is a voltage that varies according to the strength of the magnetic field passing through it. It is sensitive to the polarity of the field.

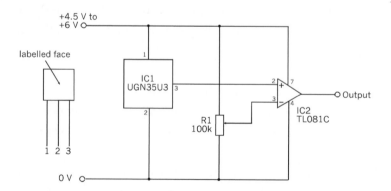

The output from the Hall effect device may be measured as a varying voltage or can be converted into a digital signal as in this circuit. The PIC's built-in comparator can be used instead of the op amp.

As a **proximity sensor** it detects the field of a small permanent magnet. For example, the *Gantry* uses small ferrite magnets as markers spaced out along the track (pp. 316-319). Each time the sensor passes close to a magnet it sends a signal to the controller. The output to the controller rises or falls, depending on the polarity of the magnet.

The device can act in the same way as a **limit switch**, to detect when a movable part of a robot, such as an arm has reached a given position. This is a variation on its use as a proximity detector. The moving part has a magnet attached to it and a Hall effect device is positioned so that the magnet comes close to it when the part is at its limit.

The device can also be used in a **tachometer**, to measure the rate of revolution of a shaft or wheel. A small magnet is mounted on the rim of the wheel and a Hall effect sensor is positioned so that the magnet passes close to it as the wheel rotates. A pulse is generated for each revolution and the pulse frequency (the number of pulses counted during a timed interval) is used to measure the rate at which the wheel is turning. If the wheel is part of the drive mechanism of the robot, the controller can calculate the robot's speed and how far it has travelled.

A **reed switch** is another magnetically sensitive device. Reed switches are often used in security systems to detect whether windows and doors are open or closed. The switch has two springy contacts that become magnetised when a magnet is nearby. They are attracted toward each other and the switch closes. They are connected to the PIC as one-bit inputs, as on p. 68.

Output circuits

Direct drive

The maximum current that can be sourced at any single terminal pin is 25 mA. The maximum current that can be sourced by the three ports at any instant is 200 mA. The same figures apply also to sinking current.

A current of 25 mA is sufficient to drive a regular LED but not the high brightness or extreme brightness types, which take 30 to 50 mA or more.

Driving an LED directly. Use the equation on p. 66 for calculating the value of the resistor.

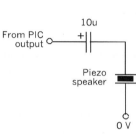

Driving a piezo speaker directly.

Transistor switches

A device that requires more than 25 mA is driven by a transistor switch. Most often the transistor is an npn bipolar junction transistor, but n-channel MOSFETs are also used. Transistor switches are used for high-power LEDs, sirens and buzzers, motors, solenoids, speakers, relays and many other devices. The devices are referred to as 'the load' in the discussion that follows.

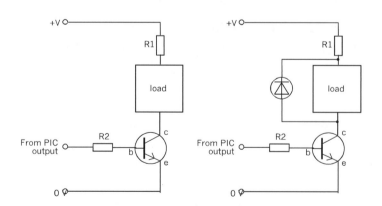

BJT switches for (left) non-inductive and (right) inductive loads. The current limiting resistor, R1, may not be necessary. The load is switched on when the input from the PIC goes high.

The point of using a transistor switch (opposite) is that a small current flowing into the base (b) of the transistor causes a current about 100 time greater to flow in at the collector (c) and out of the emitter (e).

For example, if a motor requires 500 mA to drive it, the current to the base need only be about 5 mA, which a PIC output terminal can easily supply.

When choosing a transistor it is essential to select one capable of carrying the current needed by the device. The often-used BC548 carries up to 100 mA so is not able to drive a typical motor. The BC639 carries up to 1 A, so could be the choice for driving a small DC motor.

The drawings opposite show two variations of the BJT switch. On the right there is a dioide wired in parallel with the load. This is called a protective diode because it protects the transistor from the risk of being destroyed when switching an inductive load. An inductive load is one that operates by producing a magnetic field. Examples are motors, solenoids and relays. The problem is that, when switched off, these devices induce a current that produces several hundred volts across the transistor. The diode is there to conduct this current away before it does any damage.

A transistor switch may be used to drive a device that operates at a voltage higher than the operating voltage of the PIC. For instance, if the PIC is running on 4.8 V, its output can control a motor running on 12 V. The PIC is connectd to its 4.8 V supply but the collector of the transistor is connected through the load to a supply at a higher voltage. Both circuits must be connected to the same 0 V line.

Through the transistor, the motor, running on 12 V, is controlled by the PIC, running on 4.8 V.

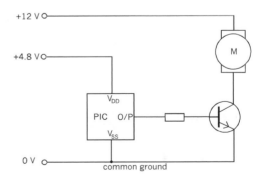

Designing a BJT switch

1) Select a transistor type that will carry the intended current and has sufficient gain.

2) When the transistor is fully on, the voltage at its collector is approximately 0.7 V. Calculate or find from data sheet the current I_{load} through load. Include a series resistor if this is too high for the load to take.

3) Voltage across base resistor is $V_{high} -0.7$, where V_{high} is the voltage of a logical high output, typically Vsupply -0.7.

4) Minimum current to switch load fully on is I_{load}/gain.

5) Base resistor is $(V_{supply} - 1.4)/(I_{load}/gain)$. Use the next higher E12 resistor.

Switches based on MOSFETs are similar, except for the fact that the drain (d) current depends on the *voltage* at the gate (g). MOSFETs have the advantage of requiring virtually no current. They are simpler to wire up as they do not need a resistor at the gate. Some have very low 'on' resistance so they deliver the maximum power to the load. They are faster than BJTs, but BJTs are usually fast enough for robot circuits.

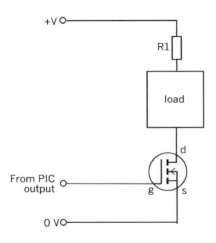

A transistor switch based on an n-channel MOSFET. The current-limiting resistor may not be required. A protective diode is necessary if the load is inductive,

Motor speed control

The simplest way to switch a motor is by a transistor switch. This turns the motor fully on or fully off, but does nothing else. There is no way of reversing the motor or regulating its speed.

The circuit illustrated below is an improvement on a simple switch. As the mechanical load on the motor varies, its speed varies. This is because the back e.m.f. generated by the motor varies, which causes the voltage across the motor to vary. The op amp acts to maintain across the motor a voltage equal to the control voltage.

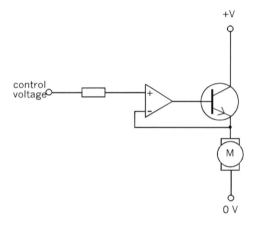

This circuit holds the speed of a motor steady even though the mechanical load on it changes.

The control voltage may be constant (say, from a fixed potential divider) to give fixed speed. For variable speed use a variable potential divider. Or a digital to analogue converter IC could be used, converting digital input from a PIC into an analogue control voltage. A simpler and better way of digital speed control is shown in the drawing overleaf.

Instead of varying the voltage, this technique powers the motor with a fixed voltage, but turns on the power for only part of the time. The power is turned on and off rapidly so that the motor runs smoothly without jerkiness.

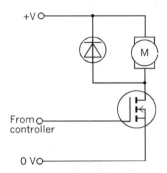

An npn BJT can be substituted for the
MOSFET.

The speed of the motor is controlled by a pulsed signal generated by the PIC. This signal has varying mark-space ratio (the ratio between pulse length and the intervals or spaces between pulses). A program is listed on pp. 252-256.

An advantage of this technique is that the motor always receives the full voltage across its terminals. It runs well at slow speeds, without stalling.

Motor direction control

There are two ways of doing this: with a transistor H-bridge, and by a relay.

The H-bridge circuit is used in most of the projects in this book. It can be based on BJTs or MOSFETS. The circuit (see opposite) comprises four transistors for controlling one motor. Two of the transistors are npn (Q1 and Q3) and two are pnp (Q2 and Q4).

There are two control inputs, A and B, which supply current to the bases of the npn transistors and turn them on when the input voltage is high. They sink current from the bases of the pnp transistors and turn them on when the input voltage is low. For instance, if A is high, Q1 is on and Q2 is off. If at the same time B is low, Q3 is off and Q4 is on. Current flows from the positive line, through Q1, through output A to the motor, through output B and Q4 to the 0 V line. The current flows through the motor from left to right in the diagram. If A is low and B is high, it flows through the motor from right to left. The H-bridge acts as a reversing switch.

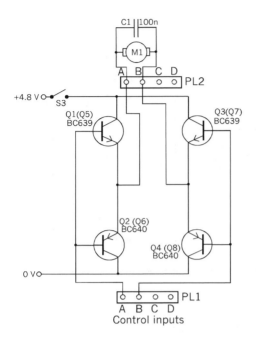

The circuit that controls the twin drive motors in the *Quester* robot. This diagram shows only the circuit for the left motor. The figures in brackets refer to the circuit for the right motor.

If A and B are both high or both low there is no way the current can flow from the positive line to the 0 V line, so the motor stops.

It may happen that the chosen motor operates on a voltage greater than 6.5 V, the maximum allowed for the PIC. The H-bridge does not work in this case. This is because a logic high from the PIC is not high enough to switch off the pnp transistors. To cope with this, a pair of npn transistors are wired as an interface between the PIC and the H-bridge. These transistors have the same higher supply voltage as the H-bridge. Their output voltage when off is close to the higher supply rail, so the pnp are properly turned off.

Interfacing the PIC to an H-bridge with higher supply voltage. Two transistors are needed, but only the A control is shown here.

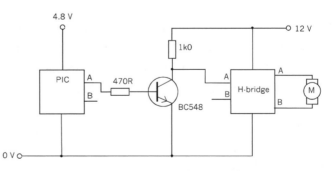

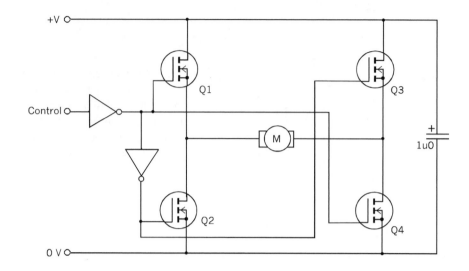

An H-bridge motor direction controller based on four n-channel
MOSFETS and two CMOS INVERT gates,

Another version of the H-bridge is shown above. This is based on four n-channel
MOSFETs. Note that in this circuit the transistors are all of the same polarity, unlike the
BJT circuit. The INVERT gates switch the transistors in pairs. Instead of INVERT gates, it
is possible to use 2-input NAND or NOR gates with their input terminals wired together.

If the control input is high, Q1 and Q4 have a low voltage at their gates and are off. Q2
and Q4 have a high gate voltage, so are on. Current flows from the positive supply,
through Q4, the motor and Q2 to the 0 V line. If the control input is low, Q1 and Q3 are
on, while Q2 and Q4 are off. The current flows though the motor in the opposite
direction.

The circuit in the diagram does not have any way of stopping the motor, but this can be
arranged by having a fifth MOSFET in series with the bridge. Connect it with its drain to
the line labelled 0 V in the diagram and with its source to 0 V. A high input to its gate
turns it on and current flows through the bridge.

As an alternative to the H-bridge as a reversing switch, a double-pole double throw relay
has some good features. The circuit being switched can be running on any voltage,
higher or lower than that of the PIC's circuit, as long as the relay contacts are rated to
accept it. Also, relays are available rated for currents of several amps.

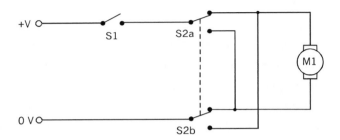

This motor reversing circuit requires two relays: S1 for on/
off and S2 for forward/reverse.

The diagram above is a typical reversing switch based on a DPDT relay. The coils of the
relays are energised by a pair of transistor switches, as seen below.

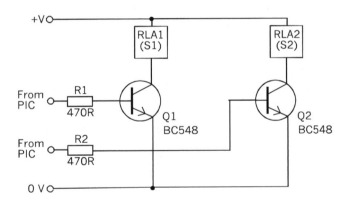

Transistor switches are used by the PIC to control the reversing
circuit.

The relays commonly used for switching motors are PCB mounting devices, often the
same dimensions as a 16-pin integrated circuit, sometimes a little larger. Most are rated to
operate on 12 V, though generally these types will operate on a lower voltage. A few
types are rated at 6 V and some at 5 V. Their contacts can take up to 1 A, which is plenty
for driving a small motor. But check the specification in the retailer's catalogue before
buying.

Servomotors

There are several types of servomotor differing in size, control method and timing requirements. A typical servomotor has three terminals. The positive and 0 V terminals are connected to the supply. The third terminal is connected to a pulse generator, which delivers a stream of pulses of fixed length. Typically, the pulse length is between 1 ms and 2 ms and the pulse frequency is 50 Hz. The pulse length determines the angle through which the shaft turns and its direction. For example, a pulse length of 1 ms makes the shaft turn to the left (anti-clockwise) as far as it will go. A 2 ms pulse makes it turn as far right (clockwise) as it can turn. A pulse of average length, 1.5 ms, makes it turn to a central position.

Often a servo is limited to turning 45° on either side of its central position, but other types can turn 90° either way. Yet others are able to turn a complete circle. Consult the data sheets before buying, and when programming the motor.

Servomotors are connected to the moving parts of a robot by a variety of horns and discs which fit on to the output shaft. Kits containing an assortment of such devices are often supplied with the motor.

Servomotors are convenient to use for steering applications as they can be put into the 'straight ahead' position at the beginning of a program by sending a train of 1.5 ms pulses for about 200 ms. This gives the motor time to reach the central position from its previous (unknown position).

Stepper motors

As their name suggests, stepper motors turn step by step. The commonest type has a step size of 7.5°, so that it takes 24 steps to make one revolution. This produces a slightly jerky motion but it is satisfactory for most purposes and the programming is simple. This description refers to unipolar stepper motors which have permanent magnets and 7.5° steps.

The motor generally has six wires, which are connected to the windings as shown in the drawing opposite.

Common
A2

A1 ——⏤⏤⏤⏤⏤——— A3

B1 ——⏤⏤⏤⏤⏤——— B3

B2
Common

The two windings of the stepper motor are tapped at their
centres.

The two taps are connected to the positive supply line. The ends of the windings are each connected to one of four transistor switches that alternate between being on or off. When the switch is off the end of the winding is at the positive supply voltage and no current flows in that half of the winding. When the switch is on, the voltage at the end of the winding is pulled down close to 0 V and the half winding is strongly energised.

The transistors at the two ends of a winding are switched oppositely so that either one or the other of the half windings is energised and the motor turns one step.

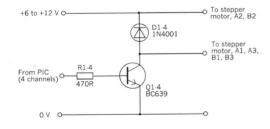

The circuit for driving the stepper motor. Four transistor switches like this are needed.

The motor is driven by turning on the transistors according to a special sequence. If 0 = off and 1 = on, the sequence is:

Q1: 1 1 0 0
Q2: 0 0 1 1
Q3: 0 1 1 0
Q4: 1 0 0 1

The transistors are turned on and off once during each sequence, but at different stages. To run the motor the PIC is programmed to generate a sequence of high pulses on four digital output channels, usually four channels of the same port. The sequence is repeated for as long as necessary. Each step is 7.5° and the sequence has four stages, so one complete sequence turns the motor 30°, so twelve times through the sequence produces one revolution. A sequence ca be halted at any step so we have positional control of the motor with a precision of 7.5°.

When the sequence of pulses is halted, the motor stops turning. But two of the transistors are still switched on and current is flowing in two of the half-windings. The shaft is held firmly by the continuing magnetic fields. The holding torque is high, of the order of several hundred gramme centimetres.

The speed of the motor is controlled by the rate at which the sequence of pulses is generated. Reversing the motor is easy — run the sequence in the reverse order.

The way to program a stepper motor is described on pp. 228-230.

Radio

Two robots that can 'talk' to each other and interact with each other open up exciting possibilities. The obvious way for them to communicate is by radio, using transmitter and receiver modules operating on the 433.92 MHz band. These modules are mass-produced for use in remote control devices, car alarms and similar applications, so their cost is minimal. Also they are intended for use in portable devices so will operate on low voltages.

There are several types of radio module available but many of them are similar to the ones descibed here. Consult the data sheet in case of doubt.

The drawings opposite show how simply they are used in conjunction with the PIC.

Transmitter and receiver modules are an effective way for two robots to communicate, or for remote control by a human. Two-way communication requires two of each of these modules.

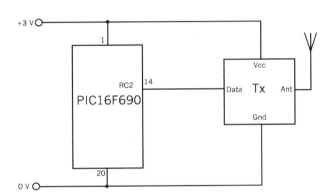

A digital signal from the PIC is sent directly to the Data terminal of the transmitter module. The drawing shows an antenna connected to the 'Ant' terminal but the module has been found to work well without an antenna for distances of several metres.

The output from the Data terminal is taken directly to a digital input channel of the PIC. Here the input signal is being copied to an LED. Again, the antenna is unnecessary for short distances, Note the higher operating voltage of the receiver module.

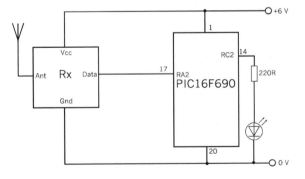

The modules require only a few milliamps. In the diagrams the PIC is run on the same supply as the transmitter or receiver. Because the transmitter operates on a lower voltage, the 6 V supply to the PIC is tapped to give 3 V for the transmitter, when the system is running both modules. Output from the PIC is fed to the transmitter through a potential divider.

Testing the circuit

Develop a routine for testing circuits at each stage of construction. Then test the completed system to check that all its parts work together as they should. Here is an outline testing routine; details depend on what kind of circuit is being tested.

1) Visual inspection: The PIC is not in its socket at this stage. The power is not switched on. Check that the circuit board wiring is correct, resistors and capacitors of the correct value, diodes and electrolytic capacitors inserted the right way round, copper strips cut where they should be cut (use a magnifier), and no solder blobs or hairs to cause short circuits (use a lens again).

2) Continuity: The PIC is not in its socket at this stage. The power is not switched on. Use the continuity checking facility of a multimeter to confirm the continuity of the 0 V line, the positive supply line(s), and any connections to off-board components.

3) Short circuits: The PIC is not in its socket at this stage. The power is not switched on. Use the continuity checker to confirm that there is no short circuit between the 0 V line and the positive supply line(s), or between adjacent pins of IC sockets, including the PIC's socket.

3) Power distribution: The PIC is not in its socket at this stage. Connect the negative lead of the multimeter to the 0 V terminal of the battery or PSU, and switch to a voltage mode. Switch on the power and check the voltage at all points that should be at supply level. If any are low, switch off immediately and look for short circuits, resistors of the wrong value, or faulty components.

4) Outputs: The PIC is not in its socket at this stage. Connect flying leads to the 0 V line and positive line. Touch these against the individual output sockets of the PIC socket to check that they operate properly, that LEDs light, motors run, and so on.

5) Inputs: The PIC is not in its socket at this stage. Use a multimeter in its voltage mode to measure voltages at the input sockets of the PIC socket. Close input (micro)switches and confirm that input voltages change correctly. Shade or illuminate light sensors and see if voltages change accordingly.

6) Testing the PIC: Switch off the power, insert the programmed PIC in its socket and switch on again. For some of the robot projects, we have written simple diagnostic programs that test the system one section at a time. These make it easier to find bugs in the system or in the programming. With a mobile robot, perch it on a support so that its wheels are in mid-air. This keeps it from running away during the test.

Suppliers

The electronic components mentioned in this Part can all be obtained from those suppliers who are geared to the hobby market. You may be lucky enough to have one or two such shops locally, and almost all operate a mail order service.

If they issue one, get a copy of your supplier's catalogue. The catalogue may be on a CD-ROM, or you may need to access it by Internet. There are dozens of websites to visit. Here are a few addresses to try:

maplin.co.uk	A well-known UK company which supplies worldwide.
jaycar.com.au	A mail order supplier of electronic components based in Australia, operates in UK and USA too. Their catalogue has a section specialising in robotics and mechatronics.
newark.com	Suppliers of a wide range of electronic components, in USA and worldwide.
crownhill.co.uk	For PICs, PICBASIC, development boards, and several interesting downloads.
ww1.microchip.com	Manufacturers of PICs, lots of data, ideas, and a link to the microchipDIRECT.com site, where a complete range of PICs are on sale by Mail Order. They also sell PICkit 1 and PICkit2.
robohoo.com	Robot arms (USA).
lynxmotion.com	Robot kits, parts, servos (USA).
budgetrobotics.com	Robot chassis kits and components (USA).

zagrorobotics.com	Robot chassis kits and other parts (USA)
parallax.com	For everything to do with the BASIC Stamp®, including Bluetooth® wireless links (USA).
imagesco.com	Supply robotic and electronic parts. Their site has an interesting selection of sensors (including a low-cost magnetic compass), articles on robotics topics, with programs in PICBASIC. They sell the PICBASIC software.
hobbyengineering.com	Robotic and electronic parts, including PICs. Wide range of kits (USA).

There are many more suppliers than we can list here. To contact some more, run a search program such as *Google* or *Yahoo!* and enter a keyword. We have discovered mountains of information by using keywords or phrases such as 'relay', 'stepper motor', and 'phototransistor'.

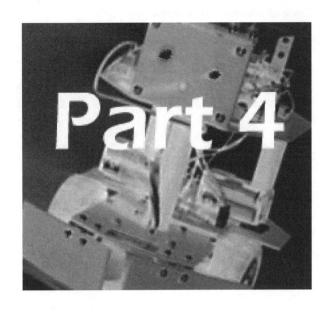

PICs in Control

4 PICs in Control

PICs and robots

The robot projects in this book are based on the PIC16F690 microcontroller, made by Microchip Technology Inc. But all PICs have a common basic architecture and electrical properties so the circuits described in this book can usually be made to work with several other types of PIC.

Programming a PIC

The PIC16F690 is especially suitable for hobby robotics because:

- It is one of the newer PICs.

- It uses the latest nanoWatt Technology. The standby current is only 1 nA when operating on a 2 V supply. At the other end of the scale, its operating current when running at 4 MHz, on its maximum supply of 5.5 V, is less than 1 mA. Low voltages and currents are ideal for driving mobile robots.

- If we opt to base timing on the built-in internal oscillator, all but two of its 20 pins are available as digital inputs and outputs. The robot can have what it needs – lots of sensors and actuators.

- It has Flash memory, ideal for programming hobby robots.

- It has 4096 words (14-bit) of program memory, which is four times that of the PIC16F84, the previous most popular PIC. It also has a generous 256 bytes of SRAM and 256 bytes of EEPROM.

- It has an assortment of on-board devices such as clocks, timers, analogue comparators, and analogue-to-digital converters useful for advanced robots.

- It is one of the cheaper PICs.

- It is programmable on the latest PICkit 2 programmer, and can be test-run on Microchip's Low Pin Count Demo Board.
- Because of their similar architecture, the mid-range PICs, such as the F690, all run on the same 35-instruction assembler language.

Assuming that the reader already knows a bit about programming in assembler (at beginner's level) this Part covers topics that are not always included in the simpler programming handbooks, especially topics useful for robotics. There are notes on using other PICs on pp. 130-132.

Hardware for programming

For the programs in this book we decided to use the PICkit 2 programmer. This is obtainable by mail order from several sources and also directly from Microchip. The website address is:

http://www. microchip.com

The site provides all kinds of useful information about the world of the PIC. Visit it and glance through the data sheet for the PIC16F690 controller. There is a link to a companion site, Microchip Direct, which supplies PICs and the PICkit 2 (and other hardware) by mail order.

The PICkit 2 comes with a USB connector. It is a small neat unit that sits comfortably on the computer bench as you program it and watch it perform. It connects directly to a small board that carries the PIC while it is being programmed. The board has a switch and some LEDs for testing input and output routines. There is space for you to build a few interfaces to the PIC.

The demo board

The PICkit 2 Low Pin Count Demo board has four LEDs driven by output channels at RC0 to RC3. These are useful when debugging a program, to indicate the state of the outputs. If your program has to send output signals through other channels, it may be possible to send them to RC0 to RC4 while the program is being developed, then change the destinations later after the program has been debugged.

The board also has a push-to-make push-button SW1 on its front left corner. This is connected to channel RA4. This pin has an external pull-up resistor so its input is normally high, but falls when SW1 is pressed. This is useful for simulating a one-bit input sensor.

For analogue input there is a variable potentiometer at front right on the board, wired between the supply and 0 V lines. The wiper of this is connected to channel RA1. Input can be swung from low (fully anticlockwise) to high (fully clockwise).

Software for programming

The most direct way of programming a PIC is first to write the program in assembler, using a simple text editing program such as *Notepad*. This, or something similar, is provided as part of the *Windows* package. The top photo opposite shows a screen-shot of an assembler listing in *Notepad*. When the file is saved, *Notepad* creates a file in text file format (.txt). When asked for the file name to save it under, type the name with the extension '.asm' (not '.txt', and not without an extension).

It is essential *not* to save the file with other formats, such a Rich Text Format. This inserts formatting control codes into the text. These codes will confuse the assembler program at the next stage.

A programmer board is usually supplied with the software that is used to turn the text file into machine code for downloading into the PIC. The software that comes with PICkit 2 includes the MPASM suite. The assembler program is called MPASMWIN, illustrated in the lower photo opposite.

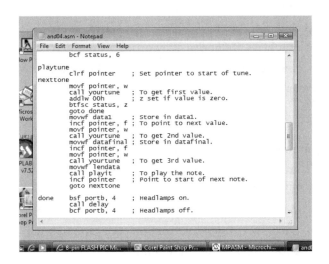

An assembler program can be written using text editor software such as *Notepad*. In this view, we see the beginning of the program for the robot of Project 6.4. Labels are placed on the extreme left of the line. Instructions are set one tabbed space in. Comments are on the right, preceded by a semi-colon. Including plenty of comments helps other people to understand the program, and help you a few months later, when you have forgotten some of the details of the programming.

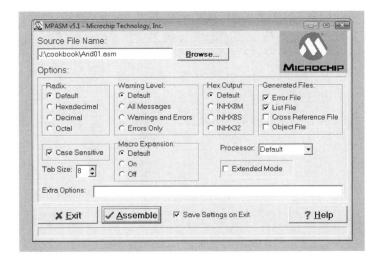

The MPASMWIN window lets you set various parameters for generating the machine code file (also known as the hex file). While assembling, it checks for errors and reports these and possibly some warnings. If there are errors they are reported in an error file (extension .err) which you can view by loading it into *Notepad*.

When using the assembler software, the only thing you have to do is to type in the source file name, select the processor type, and click on 'Assemble'. The other settings can usually be left as they are. With luck (or maybe skill), you are rewarded by a green display telling you that assembly has been successful. If you get a red display, check the error file and go back to *Notepad*. The error file has the same name as the assembler file, but with the '.err' extension.

Emulator software helps get things right more quickly. The MPLAB Integrated Development Environment, available from Microchip as part of the PICkit 2 package, is designed to do just this. It runs on your PC as if you have a PIC there and behaves just as a real PIC would behave. Its text editor does a lot of the formatting for you, including displaying the listing in several colours. Labels are in red, instruction codes are in blue bold, and so on. This makes the listing clearer to read and understand.

When you think you have got it right the IDE can simulate the running of the program. The top left window in the screen-shot below shows the listing and an arrow at the left indicates the the line that is currently being processed. The lower left window displays the contents of the File registers, while at lower right we view the Special Function registers.

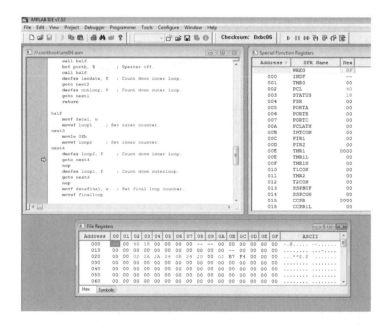

The MPLAB IDE is seen here running a simulation of a PIC16F690.

You can run the program and watch the values changing in the registers, exactly as they would in a real PIC. If you want to look more closely at a part of the program, you can go through it step-by-step. It is less confusing to write and test the program a little bit at a time. If something is wrong it is easier to spot it if only a short segment of the program is newly added.

When a part of a program or the whole program is working as you want it to, the IDE calls up programming software to transfer the assembled code to the memory of the real PIC on the programming board. Plug the PIC into its socket on the robot and the fun begins.

There is not enough space here to mention all the helpful features of the IDE. The thing to do is to download it and try it for yourself.

PICkit 2 includes programming software that can be used independently of the emulator. If you prefer, you can use *Notepad* for writing the program in assembler, *MPASMWIN* for assembling it into a hex file, and finally use the *PICkit 2* software to download the hex file into the PIC. Part 4 aims to give you an idea of what is involved when programming PICs. There are other editors and programming devices on the market, and new ones are appearing frequently. Or you may prefer to use software that operates with BASIC or C.

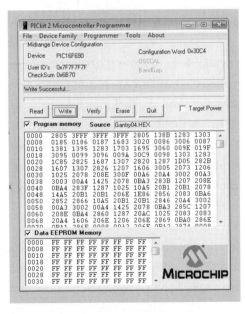

The PICkit 2 has just downloaded the source file *Gantry04.hex* into a PIC16F690. The window shows the hex code.

High-level languages

Assembler instructs the controller step by step. High-level languages, such as BASIC and C, provide the programmer with commands each of which takes the controller through many steps. This makes programming quicker and easier to follow. A good example is the WRITE command in PICBASIC. A single program line, 'WRITE 3, count', puts the value of the count variable into byte 3 of the PIC's EEPROM. Doing the same thing in assembler takes many more program lines.

A program written in a high-level language generally has shorter listings, and so takes less time to type and check. Against this is the fact that the machine code eventually compiled from a high-level language listing is usually longer than the equivalent assembler program. It takes up more memory and it takes longer to run. Assembler wins on compactness and speed of execution. A high-level language is considered by many to be easier to learn and understand.

If you decide to go for a high-level language there are several to choose from. Most come as part of an Integrated Development Environment, or IDE. An IDE runs on a PC and consists of a number of modules, such as a text editor, used when writing the programs, a compiler, to turn it into machine code, a debugger for testing the action of the code on a simulated PIC, and the software to transfer the code to the PIC's program memory.

Examples of IDEs currently available are *Proton IDE* (www.mecanique.co.uk), *SourceBoost IDE* (www.sourceboost.com) and *Microcode Studio* (www.melabs.com). They are similar in their general structure and the facilities they provide. Each has its own version of BASIC. The BASIC of *Microcode Studio* is essentially PICBASIC™, which is most like the conventional BASIC, but with additions to adapt it to programming PICs. The BASIC of the *Proton IDE* is similar to this but includes many more specialist commands. The *SourceBoost* is the most distinctive of the three, and has many features in common with C and *Java*. It is intended for professional programmers.

For those who prefer a minimalist approach, without an IDE, install PICBASIC (from microEngineering Labs (web address above) and work as described opposite. For writing and editing the programs, use a simple text editor, such as *Notepad*, which is usually provided with *Windows* software. Other editors can be used instead as long as they save files in plain text format.

Write the program in *Notepad* and save it as a text file, with the extension '.txt'. There is no need to type this as *Notepad* adds it automatically. Then (if you are not using the IDE) access DOS by clicking on the Start button at the bottom left of the screen. Next click on 'Run ...'. When the small Run window appears, type 'cmd' (without the quotes) and click on 'OK'. For future use, all of this can be set up on a short-cut icon on your desktop.

The lowest line in the command window begins with 'C:\' but there may be folder names following this. Get rid of them by typing 'CD ..' and pressing Enter. Repeat, if necessary until only 'C:\' remains, then type 'CD PBC' and press Enter to get to the PICBASIC compiler folder.

To call up the compiler, type 'PBC -p16F90 -qtxt' followed by a space and the filename of the BASIC program you have written (including the extension '.txt'. The -p option tells the compiler which PIC is being programmed. The -q option tells it to load a file with the 'txt' extension. This is explained in more detail in the PCBASIC manual.

To compile the program, simply press Enter. After a delay of a second or two a message appears stating how many words the compiled file contains. If there are errors in the text file, they are listed on screen, with their line numbers. Include blank lines when counting lines. Correct any errors and re-compile.

The final error-free file is now in the same folder, with the same name, but with the '.hex' extension. It is ready for loading into PICkit 2 or other programming board, as illustrated on p. 111.

The DOS command window, showing the line that calls up the compiler and the message that appears when an error-free program is compiled.

Flowcharts

Microcontrollers inevitably operate in small steps. They perform a long sequence of simple operations. This means that sometimes it is hard to see where the program is going, or even if it is aiming at the right goal. A flowchart gives a bird's-eye view of a program, or part of a program.

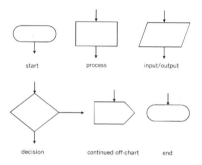

The flowchart symbols used in this book.

Flowcharts are useful at the planning stage for mapping out the main stages of a program. It may help to draw a flowchart when a program is finished, too — to see if the program really does what was intended.

Dry run

No matter what ingenious programming software is available, there are times when the only practicable way to debug a program is to do a dry run. All you need are a pencil and paper.

In a dry run you set out a table of all the registers and variables that are involved. You then go through the listing line-by-line and work out the values that are in each register. Enter these in the table and confirm that they are the right values. As an example, here is a segment of the listing for part of the random spin routine of Project 6.4 (p. 295). This is explained on p. 163.

This routine produces a new pseudo-random value in *randval* every time it is run. Repeat the routine on paper several times to confirm that the values are really appearing at random.

Listing	Contents of registers				
	bitn	bitm	randval	w	c
clrf bitn	00000000	xxxxxxxx	11011011	xxxxxxxx	x
clrf bitm		00000000			
btfsc randval, 5			(bit 5 = 0)		
bsf bitn, 0	(skip)				
btfsc randval, 6			(bit 6 = 1)		
bsf bitm, 1		00000001			
movf bitn w				00000000	
xorwf, bitm, w				00000001	
addlw '00FF'				00000000	1
rlf randval, f			10110111		

c is the carry bit of the STATUS register.
x = don't care

A dry run is sometimes the only way to get the programming just right.

Diagnostic programming

No part of a program need be permanent and LEDs are useful indicators for finding out where the PIC has got to in the listing. Use these facts to make life easier.

For instance, suppose you suspect that the PIC is not going to a particular subroutine when it should do. Add a line to the beginning of the subroutine, such as `bsf portb, 2` (or some other output channel that powers an LED). What happens when you next run the program tells you *where* to start looking for an error. If the LED does not light, it means that the subroutine is not being called. Look in the listing to find the reason for no call. If the LED lights, the calling routine is OK, but there is a fault in the subroutine itself. When you have found the fault, correct it, delete the added line, and re-assemble.

The PIC 16F690

Programming a PIC mainly consists of writing data into arrays of registers. There are two kinds of register:

- **Special Function Registers:** Hold data related to the input and output ports, comparators, timers, other functions, and the general running of the controller. More about some of these in the pages that follow.
- **General Purpose Registers:** This is where the program writer stores data and processes it.

Data in these registers is stored as bytes, and is lost when the power is switched off. In the Special Function Registers the data takes initial default values when power is switched on. Often the default value is what you want, and you need to access the register only if you want to set it to a different value, or to read data from it.

Pins and ports

The F690 version that we describe here has a 20-pin double-in-line package (opposite). Other packages are obtainable, such as surface-mount devices, for example.

All pins except 1 and 20 are available for use as input/output pins. Those of Ports A and B can be individually set to have built-in pull-ups when configured as inputs. Also, as inputs, they can be programmed to cause interrupts when the input signal changes.

Certain of the pins, though usable for simple input or output, can be programmed to have special functions. For example, pin 4, RA3, can be programmed to act as a master clear input which, when made low, resets the PIC to the start of its program.

Another example is pin 10, RB7. If you are using the USART (pp. 124-126), this pin must be used for data that is being sent to another PIC by line or radio. There is more about these special pin functions later in this Part.

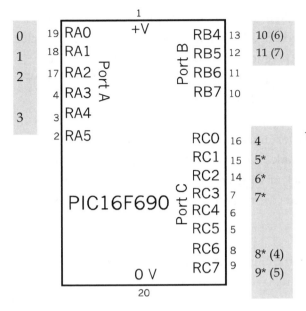

Ports and pin numbers of the PIC16F690 controller. The channels of each port are numbered from 0 to 7, from the least significant to the most significant bit. Port A has only six channels. Port B has channels for only the four most significant bits.

The figures on grey are the digital input channels AN0 to AN11. For AN0 to AN7 these figures are also the bit numbers in the ANSEL register. For AN8 to AN11, the bit numbers in the ANSELH register are given in brackets,

The channels marked with an asterix are also availilable for input to the comparators.

Bits

Terms used in all the descriptions are defined as follows. The bits of a byte are numbered from right (bit 0, the least significant bit or LSB) to left (bit 7, the most significant bit or MSB).

When we refer to a bit in a given register, we give the register name followed by the bit number in angle brackets. For example, the bit called RB5 (bit 5 of PORTB register) is referred to as RB<5>. This is the convention used in the PIC data sheets.

A range of bits is defined by quoting its most and least significant bits, in that order. For example the lower 4 bits of PORTB are RB<3:0>.

Special function registers

These registers control all aspects of the operation of the PIC. They can be written to or read from as a whole byte of data or as individual bits. In some cases the bits are read-only or write-only. These, together with the 8-bit **working register** (known as w), are where the action is.

The Special Function registers are located in four areas of memory, of which we can access only one at a time. These areas are named Bank 0 to Bank 3. Some registers, such as the STATUS Register, appear in the same position on all four banks. Others appear in only one or two of the banks. For example, TRISA (which determines which bits of Port A are inputs and which are outputs) appears only on Banks 0 and 3.

The bank that is currently accessible is determined by the values of bits 5 and 6 of the STATUS register. For example to select Bank 2 set bits <6:5> to '10', or binary 2. It is essential when programming to keep track of which is the current Bank. For normal working we operate in Bank 0, which holds the most often used registers such as STATUS and Port registers.

As an example of Bank switching, the first thing in most programs is to initialise inputs and outputs. At power-up all pins are inputs, so to flash an LED we have to make its pin an output. The LED might be connected between pin 16 and the 0V line. Pin 16 is the RC0 pin, connected to bit <0> of Port C.

Assume we are working in Bank 0, with STATUS<6:5> = '00'. To make RC0 an output we switch over to Bank 1 by setting STATUS<5>. Then we make RC0 an output by clearing the corresponding bit in TRISC. That done, we return to Bank 0 by clearing bit 5 in the STATUS register.

The program listing for this operation is:

```
bsf status, 5      ; Bank 1.
bsf trisc, 0       ; Make RC0 an output.
bcf status, 5      ; Back to Bank 0.
```

There are many more instances of Bank switching in the listings in this Part and in Part 6.

Configuration word

This is a 14-bit word that is stored at a special address in memory and which sets up the way in which the controller is to operate. One of the first things to be done by a program is to define this word (p. 135).

Without going into details about several of the functions, the word used for the programs listed in this book is made up as follows:

<13:12>   Not implemented, set to 11.

<11:6>   Functions enabled or disabled for simplest operating, set to 000011.

<5>   Make pin 3 (RA4) a digital input, set to 0.

<4>   Enable power-up timer, set to 0.

<3>   Disable watchdog timer, set to 0.

<2:0>   Select INTOSCIO, internal oscillator with pin 2 (RA5) a normal input/output pin, not an output pin for the oscillator set to 100.

Putting all these settings together gives:

11 0000 1100 0100

In hexadecimal notation (hex for short), this is:

30C4

Using other settings may make very little difference to the way the program runs, but we are assuming in our programs that bit <5> is '0' and the internal oscillator is selected.

Ports

The ports connect the PIC to the outside world. Each port is represented by a Special Function Register on banks 0 and 2. These are labelled PORTA, PORTB, and PORTC. Although these registers are a byte each, ports A and B do not have all the channels implemented (see diagram, p. 117).

The same addresses on Banks 1 and 3 refer to the tristate registers of the three ports. These are labelled TRISA, TRISB, and TRISC. These registers control whether the channels are for input or output. Setting the corresponding bit to '1' makes the channel an input. A '0' makes it an output. This is easy to remember if you think of them as 1inputs and 0utputs.

At switch-on all channels are set as inputs so if you want them all to be inputs you need take no further action. However in the F690 the channels are automatically set to be *analogue* inputs. If you want digital inputs, program them as described on p. 139.

Special functions

Comparators

The 16F690 has two voltage comparators, each of which has its own control and register. For comparator 1 (C1), this register is called CM1CON0, at address 119h in Bank 2. We will look at programming C1. Programming C2 is similar. The comparator has the usual properties of a comparator circuit. It has two inputs, C1VP (positive, or non-inverting input) and C1VN (negative, or inverting input). The non-inverting input is supplied either through the RA0 pin (pin 19) or from an internal voltage reference which can be set to a number of levels. The inverting input is supplied through one of four analogue channels 7, 6, 5 or 1 (see p. 117).

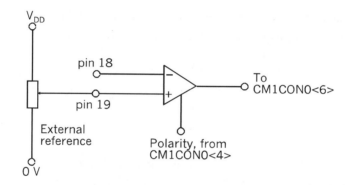

A PIC comparator visualised as a schematic. If the polarity bit is 0 (non-inverted polarity), the output is high when the input at pin 18 is less than the reference voltage.

The comparator has a single output C1OUT which can be read as a voltage at pin 17 or as the value of bit 6 of the CM1CON0 register. There is a polarity control bit should you need to invert the state of the output. The reference voltage is set by using the voltage reference control register, VRCON, address 118h. The procedure is explained opposite.

As an example, this is how to set up CM1 with its non-inverting input through the RA0 pin (pin 19), its inverting input through RA1 (pin 18) and its output sent to bit 6 of CMCON0. Polarity is to be non-inverted. Working with a regular hardware comparator IC such as the LM311, the connections would be as in the diagram above.

At reset, all bits in CM1CON0 are set to 0. The settings we need are:

> <7> = 1, to enable the comparator.
>
> <6> = 0, this is the output bit.
>
> <5> = 0, output to CM1CON0<6>, not to pin 17.
>
> <4> = 0, for non-inverted polarity.
>
> <3> = 0 (no function).
>
> <2> = 1 , use internal reference.
>
> <1:0> = 00, input at pin 18.

Settings of <1:0> for accepting input at other input pins are 01 (pin 15), 10 (pin 14) and 11 (pin 7). To set Comparator 1 as above, the byte is 10000100, or 84 in hex, which is to be stored at 119h.

Similarly for Comparator 2; the byte 10000101, or 85h, when stored at 11Ah sets that comparator to accept input at pin 15.

As we are using the internal reference this must be programmed by setting the VRCON register. This can provide either a 0.6 V constant reference or a variable reference in the range 0 V to $0.71875 \times V_{DD}$. The reference voltage is applied to the inverting input.

To set up both the comparators for the variable internal reference, the bits in VRCON are as follows:

> <7> = 1, to use variable reference for Comparator 1.
>
> <6> = 1, to use variable reference for Comparator 2.
>
> <5> = 0, to select low range (or = 1 for high range).
>
> <4> = 0, constant reference disabled.
>
> <3:0>, four bits to select the range of the variable reference. In the low range the reference is (this value) / 24 × V_{DD}. In the high range it is $V_{DD}/4$ + (this value) / 32 × V_{DD}.

For example, the word 11001100 (CCh) stored at 118h sets both comparators to use the variable reference in its low range. The value of bits <3:0> is 1100 (decimal 12), so the reference is $12/24 \times V_{DD}$, that is half the supply voltage.

The output of the comparators is either 0 or 1 and is readable at bit <6> of CM1CON0 and CM2CON0. Both outputs can be read at the same time as bits <7> and <6> of a third register, CM2CON1, at 11Bh.

As well as the registers directly involved with the comparators we must also enable the inputs for analogue voltages. This is done through the analogue select register ANSEL, at 11Eh.

Pin 18 is the AN1 input and pin 15 is the AN5 input so we set bits <1> and <5> to 1. The byte is 00100010, or 22h.

The registers concerned with the comparators are all in Bank2, so the routine for setting up Comparators 1 and 2 with a reference of half the supply voltage is:

```
bsf status, 6         ; Page 2.
movlw 22h             ; Define AN1 (pin 18) and AN5
movwf ansel           ;    (pin 15) as analogue inputs.
movlw 84h             ; Set up Comparator 1.
movwf cm1con0
movlw 85h             ; Set up Comparator 2.
movwf cm2con0
movlw CCh             ; Set up reference voltage for both
movwf vrcon           ;    comparators.
bcf status, 6         ; Back to page 0.
```

By the way, all listings in Part 4 assume that the addresses of the registers have been declared by using EQU directives at the beginning of the program (p. 135), or that they are defined by the compiler program. If the controller is already in Page 1 mode when this routine is run, bit 5 of STATUS must be cleared first.

To read the output of Comparator 1 use these lines:

```
bsf status, 6         ; Page 2.
movlw 40h             ; Bit 6 of this byte is '1'.
andwf cm1con0, w      ; Read the output bit, result in w.
bcf status, 6         ; Back to Page 0.
```

The ANDing operation sets or clears the zero bit of the STATUS register, which may then be tested in the usual way and suitable action taken.

Analogue to digital converters

Analogue input is fed to the converter through any one of 14 channels. For this discussion we will refer only to channel AN0, which is at pin 19. At power-on, all the PIC's output channels are automatically set as analogue inputs. The analogue selection register ANSEL, which is in Bank 2, is used to define which are to be reset as digital inputs. If bit <0> is '1' this makes AN0 (pin 19), an analogue input. At the same time it disables the weak pull-up and the interrupt-on-change function.

The next step is to select the input channel and other functions, using ADCON0. The setting used in this book is:

<7> = 0. result of conversion left justified (see below).

<6> = 0, use supply voltage as reference.

<5:2> = 0000, select AN0 as input channel (pin 19). These bits can take other hex values for channels AN0 to AN11 (1011).

<1> = 1. Set this bit to start a conversion, then read it; it stays 1 while the conversion is in progress, and goes low when it is completed. This is the GO/DONE bit.

<0> = 1. This enables the converter, 0 turns it off. In summary, setting <0> sets up the converter on channel AN0 and setting <1> starts the conversion.

The final step is to select the clock rate for the conversion, in the ADCON1 register bits <6:4> do this. In our circuits there is no hurry so we can use the slowest rate, and in this case can leave these bits clear, as on reset. In other words, nothing to do.

The output is a 10-bit value and so needs two registers to hold it. These are ADRESH and ADRESL. With left justification selected (see above), the 8 most significant bits appear in ADRESH and the 2 least significant bits in bits <7:6> of ADRESL. Bits <5:0> of ADRESL read as 000000. This is how it looks:

```
        ADRESH          ADRESL
     X X X X X X X X   X X 0 0 0 0 0 0
      ——— 10-bit result ———
```

If high precision is not required (or is not possible), the content of ADRESL can be ignored. Ignoring this gives 8-bit precision (1 in 256, or approximately 0.4%) which is better than most of our sensors can provide.

Using channel AN0 and reading a low-precision result turns out to be relatively simple as nothing to do to certain of the registers. The only registers to be set are ANSEL in Bank 2 and ADCON0 in Bank 0. The result is read in ADRESH, which is also in Bank 0. Here is a simple routine:

```
        bcf status, 5
        bsf status, 6      ; Bank 2
        bsf ansel, 0       ; Select channel AN0
        bcf ststus, 6      ; Back to Bank 0
        bsf adcon0, 0      ; Turn on converter
        bsf adcon0, 1      ; Start conversion
wait:
        btssc adcon0, 1    ; If conversion complete
        goto wait          ; If not complete
        movf adresh, w     ; Read 8-bit result into
                           ;  working register.
```

The result in the working register is ready for processing.

Data transmission with the USART

The 690 has an Enhanced Universal Synchronous Asynchronous Receiver Transmitter, or USART for short. We use this to take a byte of data and transmit it serially (that is, one bit at a time) to the USART of another PIC. The second USART receives the serial data bit by bit and assembles it into a byte that can be read from a register. Transmission can be by a single wire (if the 0V line is common to both PICs) or by a radio link.

The output from the USART is at RB7 (pin 10) and the input at RB5 (pin 12). The USART can operate in both synchronous and asynchronous modes but we describe only the asynchronous mode. In this mode, one USART waits for an indefinite period to receive the data signal from the other USART.

As a transmitter, the USART is controlled by the TXSTA register, in Bank 1. The settings we use are:

<7> = 0, not applicable for asynchronous mode.
<6> = 0, 8-bit transmission.

<5> = 1, enable tranmission.

<4> = 0, asynchronous mode.

<3> = 0, disable Break Character bit.

<2> = 0, low speed.

<1> this bit is '1' if transmit shift register is empty and '0' if it is full.

<0> = 0, the ninth bit, when used.

All we have to do to transmit a byte is to place it in the data register TXREG and set bit <5> of TXSTA. There is no need to set the Baud Rate Control register (BAUDCTL), as we use the default asynchronous mode.

After a short delay to allow data (usually only a single byte) to be transmitted, the data is read from RCREG in the other PIC. To set up the USART as a receiver, the RCSTA register is loaded as follows:

<7> = 1, enables serial port.

<6> = 0, 8-bit reception.

<5> = 0, not applicable.

<4> = 1, enables receiver.

<3:0> = 0, not applicable.

The following is a routine for configuring the USART:

```
bsf status, 5          ; Bank 1.
bsf trisb, 7           ; RB7 as input (automatically changed
                       ; to output).
bsf trisb, 5           ; RB5 as input.
clrf txsta             ; Set up, but not enable transmitter.
bcf status, 5          ; Back to Bank 0.
bsf rcsta, 7           ; Set up but not enable receiver.
```

The USART is now ready for action except that it is not enabled. Immediately or whenever we want to use it we switch it on by:

```
bsf status, 5          ; Bank 1.
bsf txsta, 5           ; Enable transmitter.
bcf status, 5          ; Bank 0.
```

A subroutine for transmitting a byte is:

```
        movlw XXh           ; XX is the byte, in w.
    transmit
        btfss pir1, 4       ; If register clear for use.
        goto transmit
        movwf txreg         ; To send the  byte.
        return
```

This assumes that the transmitting PIC knows that the other PIC is ready to receive. It may previously have sent a 'ready to receive' byte.

To receive a byte, the code is:

```
    receive
        bcf status, z       ; Reset zero flag.
        btfss pir1, 5       ; New data received?
        return              ; No, try later.
        btfss rcsta, 2      ; Test for framing error.
        goto noferr         ; No framing error.
        movf rcreg, w       ; Byte to w.
        bcf status, z       ; Zero flag clear - error.
        goto orerr          ; to check overrun flag.
    noferr
        movf rcreg, w       ; Byte to w.
        bsf status, z       ; Set zero flag to show valid byte.
    orerr   btfss rcsta, 1  ; Test for overrun error.
        return              ; No overrun error.
        bcf rcsta, 4        ; Clear continuous receive.
        bsf rcsta, 4        ; Enable it (clears overrun flag).
        return
```

The subroutine returns with z = 1 if there is a new valid byte in w. Call the above subroutine, check z, and either use the byte that is in w, or call again until a valid byte is received.

The routines given above show you how to set up the USART, how to send and how to receive data. The other thing to think about is how to write these routines into a program so that data can be exchanged with another controller while the program continues to run.

Data memory

The 16F690 has 256 bytes of EEPROM data memory that can be written to for long-term storage of data. Unlike the ordinary SRAM used as temporary storage, the data in this memory remains for 40 years or more or until it is overwritten.

Most important is the fact that it is not lost when the power is switched off. For this reason it is useful in robots that can be taught or in other ways can learn their best responses to given situations. They do not forget what they have learnt. They can carry their acquired knowledge over from one session to the next.

The data memory is addressed in the range 0 to 0FFh. Reading data is simpler than writing it. A routine for reading data from EEPROM is as follows:

```
bsf status, 6      ; Bank 2.
bcf status, 5
movlw address-to-read
movwf eeadr        ; Address to eeadr register.
bsf status, 5      ; Bank 3
bcf eecon1, 7      ; Access data memory.
bcf eecon1, 0      ; Read data.
bcf status, 6      ; Bank 2.
movf eedat, w      ; Data to w.
bcf  status, 5     ; Bank 0.
```

Writing data includes a sequence of instructions that must be followed exactly:

```
bsf status, 6         ; Bank 2.
bcf status, 5
movlw address-to-write-to
movwf eeadr           ; Address to eeadr register.
movlw data-to-write
movwf eedat           ; Data to eedat register.
bsf  status, 5        ; Bank 3.
bcf eecon1, 7         ; Access data memory.
bsf eecon1, 2         ; Enable writing.
bcf intcon, 7         ; Disable interrupts. Required code begins here.
movlw 055h
```

```
        movwf eecon2          ; Write 55h.
        movlw 0aah
        movfw eecon2          ; Write aah.
        bsf eecon1, 1         ; Write data.   Required code ends here.
writing
        btfsc eecon1, 1       ; Becomes 0 when data is completely written.
        goto writing          ; Until data written.
        bcf eecon1, 2         ; Disable writing.
        bcf status, 5         ; Bank 0.
        bcf status, 6
```

The PIR1 and PIR2 registers

The two Peripheral Interrupt Request registers are helpful when you need to know the current state of the PIC's built-in peripheral devices. These are the analogue-to-digital converters, the USART receiver and transmitter, and the comparators. On being interrupted, the PIC goes to the interrupt service routine (ISR) and reads data from each of several sources to find out the reason for the interrupt. But reading from these devices before they have completed their current processing gives a false result. We must let them finish what they are doing, and this is where the PIR registers are important.

PIR1 has three flags indicating the readiness or otherwise of the currently operating A-to-D converter, the receiver and the transmitter. Bit <6>, known as ADIF (Analogue-to-Digital Interrupt Flag) goes high when the curent conversion is complete. If you read this bit and find it is 0, you must try again later. Bit <5>, or RCIF, goes high when the receiver buffer is full, a 0 indicates that data is still accumulating in the buffer, so try again later. This flag is automatically cleared when you read the buffer data held in RCREG (p. 126).

Bit <4>, or TXIF, goes high when the transmitter buffer is empty and is waiting to receive more data to be transmitted. Otherwise it is 0, indicating that the buffer is full, and waiting to transmit it.

PIR2 has two flags that indicate that the output of a comparator has changed. Bit <6>, or C2IF, goes high when the output of Comparator 2 has changed, while bit <5>, or C1IF, goes high at a change in Comparator 1.

The bits remain set until you reset them to 0 by a bit clear operation. This means that we do not have to read the bit as soon as a comparison has been made. We can access the bits at any later time. However, if you want to use these bits to monitor further comparisons you must clear the bits to 0 at some stage before then.

All these flag bits are set when the events occur, whether or not interrupts are enabled, so you can use them even if you are not using interrupts. They tell you when it is safe to read data from these peripherals. But if you switch interrupts on later in the program, remember to reset the flags before you do it.

The INTCON register

This register controls the processing of interrupts. The bits of interest are:

<7> GIE 1 = interrupts enabled, 0 = interrupts disabled. This bit is used to switch all interrupts on or off with a single command.

<6> PEIE 1 = enable peripheral interrupts, 0 = disable these interrupts.

<5> TOIE 1 = enable timer TR0 overflow interrupts, 0 = disable.

<4> INT 1 = enable INT interrupt, 0 = disable. The INT interrupt is an external interrupt occurring when RA2 changes. The direction of change that brings about an interrupt depends on INTE, which is bit <6> of the OPTION register: 1 = interrupt on rising edge, 0 = interrupt on falling edge. By default INTE is 1.

<3> RABIE 1 = enable interrupt on change in Port A and Port B. You can select which bits of Port A have the interrupt on change feature by setting one or more bits in the IOCA register.

The lower three bits are flags:

<2> T0IF is set when the timer TM0 overflows.

<1> INTF is set when there is an INT interrupt (see above).

<0> RABIF is set when there is an interrupt on change in Ports A and B.

Like the flags in the PIR registers, the flags remain set until you clear them.

Other PICs

Because PICs share the same intruction set, programs written for one type of PIC can often be run on another type. The program may need to be modified for various reasons but this usually presents few problems.

The table below describes three popular mid-range PICs that can run at least some of our robotic programs without too many amendments. All the types are supported by MPASM, MPLAB IDE, PICkit 2 and the PICBASIC compiler.

Device number	16F84A	16F628A	16F88
Pins	18	18	18
Ports (bits)	A (5), B (8)	A (8), B (8)	A (8), B (8)
General purpose RAM (bytes)	68	224	368
Program ROM (words)	1 k	2 k	4 k
EEPROM (bytes)	64	128	256
On-chip peripherals	1 timer	3 timers 2 comparators Capture/compare Internal oscillator	2 timers AD converter 2 comparators Capture/compare Internal oscillator USART

Popular PICs compared.

The major change is in the controller boards. All the PICs in the table have 18 pins and the layout of power terminals is different, as can be seen in the pinout diagrams below. Taking the PICs one at a time, the other important differences are:

16F84A: Has no internal oscillator so an external one is required. The diagram shows an RC oscillator. The programs in this book assume that the 16F690 is running on its internal clock rate of 4 Mz. For programs to run at the same speed on the 16F84A the external oscillator must have a 4 MHz crystal. This is the maximum rate for the 16F84A.

16F628A: This has an internal oscillator running at 4 MHz by default. Using this frees two more pins for input/output, bringing the total to 16. This PIC has more peripheral devices than the previous one, so has become very popular. The diagram below shows its pinout.

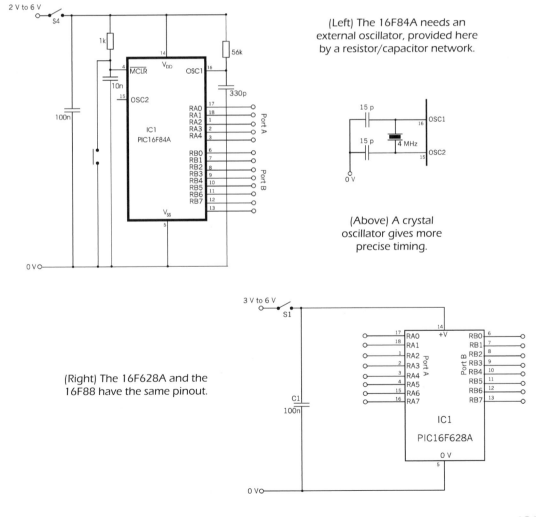

(Left) The 16F84A needs an external oscillator, provided here by a resistor/capacitor network.

(Above) A crystal oscillator gives more precise timing.

(Right) The 16F628A and the 16F88 have the same pinout.

16F88: This controller scores over the others (but not over the 16F690) by having a larger program memory. It also has an analogue-to-digital converter.

The increase in the number and capabilities of the peripherals inevitably leads to more multiplexing of the input and output channels. Setting up the peripherals becomes more complex because there are more options to choose from. Although straightforward operations such as one-bit input and output remain unaffected, control of the peripherals involves more registers.

There are also differences in the properties of the I/O channels — whether they are digital or analogue, whether they have weak pull-ups, whether they have interrupt-on-change. For further details download the data sheet from Microchip.

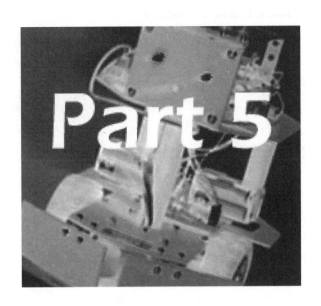

PIC Programming

Program segments

As you can see if you look at the programming examples of the projects, there are program routines that we use over and over again. Some may differ in detail but essentially they are repeats. It makes more sense to save these as separate text files, so that you can load them and put them into your current project. They may need a little editing here and there but this is nothing compared with the effort needed to type them in from scratch every time you need them.

Not only does it save typing effort, but it reduces the risk of typing errors too.

The program segments listed in this Part are a selection of the most often needed routines. The listing opposite begins with a title frame. It is not strictly necessary, but in a few months' time you may have forgotten what the program is supposed to do. Every line of the title frame must begin with a semi-colon so that it is ignored by the assembler.

The first active line of the listing tells the assembler which PIC you are using. Then comes the configuration directive which sets up the key features of the way the PIC is to operate.

With development software such as MPLAB IDE, it is not necessary to type the list of equates which comes next. All this data s contained in a linker file which is automatically loaded when the assembler is run. You will still need to type in equates for the labels of the registers that your program uses. An example is delay0, used in the *delay* subroutine (see pp. 285-286).

When working simply with an assembler such as MPASM a list of equates is required. The list opposite is a long one because there are many that you may need to use. Not that you are likely to need them all in any one program. We made the list comprehensive on the basis that it is preferable to type in the lot once and for all, and get the particulars right. In any given program you can, if you want to, simply delete those that are not used.

```
;**********************************************************
;
;     Filename: Leader.asm
;
;     The directives etc for setting up a typical program.
;
;**********************************************************

      list p=16F690
      __config 0x30c4

; Bank0
      pcl             equ 02h
      status          equ 03h
      porta           equ 05h
      portb           equ 06h
      portc           equ 07h
      intcon          equ 0bh
      pir1            equ 0ch
      pir2            equ 0dh
      rcsta           equ 18h
      txreg           equ 19h
      rcreg           equ 1ah
      adresh          equ 1eh
      adcon0          equ 1fh
; Bank1
      option_reg      equ H'01'
      trisa           equ H'05'
      trisb           equ H'06'
      trisc           equ H'07'
      ioca            equ H'16'
      txsta           equ H'18'
      adresl          equ H'1e'
      adcon1          equ H'1f'
 Bank2
      eedat           equ 0ch
      eeadr           equ 0dh
      vrcon           equ 18h
      cm1con0         equ 19h
      cm2con0         equ 1ah
      cm1con1         equ 1bh
      ansel           equ 1eh
      anselh          equ 1fh
; Bank3
      eecon1          equ 0ch
```

Leader listing (continued overleaf).

```
; Destination codes and zero-bit code

    w           equ 00h
    f           equ 01h
    z           equ 02h

; Labels
            delay0          equ 20h
            delay1          equ 21h

    goto start
    org 04h
    goto start

start
    goto $


; Subroutines

delay
    decfsz delay0, f
    goto delay
    decfsz delay1, f
    goto delay
    return


    end
```

Leader (continued from p. 135).

After the equates for registers, there are equates for the single codes, w, f, and z. These are optional but well worth including. Some of the PIC's instructions are of the form 'do this and put the result there'. 'There' is the working register (w) or the register currently being operated on (f). For example, to decrement the value in a register named count and put the result in the working register, the instruction is:

```
                decf count, 0
```

This leaves count unchanged. To put the result in count, without affecting the working register the code is:

```
                decf count, 1
```

Using the equate equivalents, w and f instead of 0 and 1, these instructions become:

```
decf count, w
```

and

```
decf count, f
```

which makes it much clearer (to the programmer) what is happening.

After the equates there is a list of labels for the general purpose registers that are used by the program. These start at address 20h and there are up to 96 of them, finishing with address 7fh. The first 80 of these (20h to 60h) are accessible in Banks 0 to 2, but the last 16 addresses are only in Bank 0.

In this listing there are only two labelled registers required for this delay subroutine.

Next the start address is listed in the conventional way, so that there are four lines reserved for a jump to the interrupt service routine (if there is one) in the event of an interrupt.

At the `start` label, the statement 'goto $' sends the processor back to `start`, so producing a continuous loop. This may not work with some assemblers, in which case substitute 'goto start'. Either version is a temporary action which prevents the program from running on to the delay subroutine. The `start` label is where you begin typing in your own program.

The leader includes delay subroutines. These are almost essential in any program, so why not include them here? Normally, subroutines go at the end of a listing (though some people prefer to have them near the beginning). The order makes no difference to the way the program runs.

One essential line — the program must end with an `end` directive.

Inputs and outputs

Programs usually begin by clearing the ports and setting each channel as either an input or an output. We also need to think about which input channels are to be digital and which analogue. If a channel is a digital input, is it to have weak pull-ups?

```
start
      bcf intcon, 7        ; Disable interrupts.

      bsf status, 5        ; Bank1.
      movlw 3eh            ; <1:5> inputs, rest outputs,
      movwf trisa          ;    for Port A.
      movlw 20h            ; <5> input, rest outputs,
      movwf trisb          ;    for Port B.
      clrf trisc           ; All outputs.
      movlw 06h            ; <2:1> interrupt on change,
      movwf ioca           ;    port Port A.
      bcf status, 5        ; Bank2.
      bsf status, 6
      clrf ansel           ; For digital I/O.

      bcf status, 6        ; Bank0.
      clrf porta
      clrf portb
      clrf portc
```

A typical setting up routine.

Now we step warily through the various banks of Special Function registers. Incidentally, even though a linker file may have eliminated the need to type the equates, the list of registers on p. 135 is handy as a reminder of which bank each register is in. INTCON is in all four banks, but some of the registers are not. It is important to be aimed at the right bank when setting a register.

The bank is selected by setting or clearing bits 6 and 5 of the STATUS register. On power-up, these bits are both '0', selecting Bank 0, which is the default.

Setting STATUS <5> changes STATUS <6:5> to '01' which selects Bank 1. Here are found the tristate registers, which make each channel of a port an input or an output. On power-up all bits are set (= 1), so all channels are inputs. By resetting a bit to 0, the corresponding channel becomes an output. In this example, all bits of Port A are to be inputs, so leave TRISA in its default state. In Port B, all bits are outputs, except for bit 5. The bits must therefore be 0010000, or 20h. This value is put into the working register, w, by `movlw 20h`, and then moved to TRISB by `movwf trisb`. Port C channels are all to be outputs. Simply clear all bits at one go by `clrf portc`.

A weak pull-up is the equivalent to a high-value resistor connected between an input terminal and the positive supply. It makes the input read as logic high, unless the external circuit takes it low enough to read as logic low. This feature replaces external pull-up resistors, though these may be necessary if the input signal is noisy with sharp voltage spikes.

Weak pull-ups are available on Ports A and B, but not on Port C. In the example opposite they are disabled by default because the global enabling bit 7 of the option register (OPTION_REG) is set. If pull-ups are required, clear this bit, to enable *all* the WPUs by default, then clear the corresponding bits in registers WPUA and WPUB to disable the WPUs that are not required.

Port A in this example has all channels as inputs and connected to switches that take the channel to 0 V when closed. The default condition is accepted for Port A. In Port B, all channels are outputs except for bit 5.

In this project there are two microswitches that need to be monitored for interrupts. These are connected to bit 1 and 2 of Port A. Therefore the code setting is 00000110, or 06h. This value is put in the Interrupt On Change register, IOCA. The interrupt on change setting does not take effect until the Global Interrupt Enable bit (GIE) is set. This could be done straight away, by setting the Global Interrupt Enable bit:

```
bsf intcon, 7
```

In the example we do not want interrupts just yet, so we leave GIE as 0.

All inputs in this application are to be digital but, in the 16F690, all the channels that can act as inputs for the AD converters are analogue channels by default. If we want them to be general-purpose digital inputs, we have to clear the corresponding bits in the ANSEL and ANSELH registers. In this example, we do not use AD converters. Select Bank 2 and clear the whole ANSEL register, making all channels that are inputs into digital inputs. There is no need to do anything to ANSELH as the channels are all outputs except for RB5, which is under the control of the USART.

That completes the initial setting up. Return to Bank 0, ready for the main program but, before doing this, it is a safety precaution to clear all the output channels, unless there is a reason for not doing so.

Mode select routine

The memory of a PIC is large enough to accommodate several different programs — unless they happen to be blockbusters. For this reason it is convenient to have a number of programs in a single PIC chip and to be able to select any one of these at run time.

The first program lines after the 'start' label usually initialise the ports and set any options that are in force for the whole program. Immediately after this comes the mode select routine. Leave it out if there is only one program on the chip.

Below is a typical mode select routine. In this example, the two mode select switches are connected to RC0 and RC1, as in the *Quester* project.

```
        btfsc portc, 1        ;Test mode select bit 1.
        goto bit1hi
        btfsc portc, 0        ;Test mode select bit 0.
        goto mode2
        goto mode1

bit1hi
        btfsc portc, 0        ;Test mode select bit 0.
        goto mode4
        goto mode3
```

An example of a mode select routine.

The program then branches to the four sub-programs (modes), which are listed one after the other, each beginning with a modeX label.

Opposite is the flowchart of this routine. It is a good example of the use of branching instructions. There are three two-way branch points, which lead to the four modes.

It is essential that the subprograms are entirely separate. It must not be possible for the PIC to run on from one mode to the next in the listing. However, sub-programs can share the same sub-routines, such as delays.

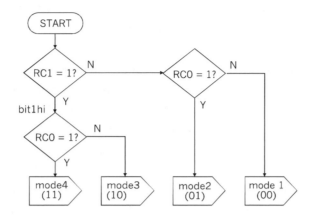

Flowchart of the Mode Select routine.

Branching instructions

The PIC instruction set has two branching instuctions, one of which was used in the Mode Select routine (opposite). The other instruction is the converse of this, test a bit and skip the next instruction if it is set (btfss). These two are the only instructions for implementing a decision box in the flowchart.

In the listing opposite we used the instruction directly. The PIC tests a bit and acts accordingly. Often the instruction tests a bit that has been set or cleared as the result of a previous operation. Very often the bit tested in the zero bit, bit <2> in the STATUS register, which is why we set up a special equate for 'z'.

There are 15 instructions that affect the zero bit. These are the operations that can have a zero result. For example, decf decrements a given register and, if the result is 00h, sets the zero flag to 1. If the result is not 00h, the flag is cleared to 0. We use the state of the flag to branch one way if z = 1 and another way if z = 0. The routine might look like this:

```
decf count, f       ; Decrement register count and put result
                    ; in count.
btfsc status, z     ; Test zero bit and skip next instruction
                    ; if it is clear.
goto finish         ; Go to the finish label if count is
                    ; 00h.
```

... continue with the program.

Other instructions that result in a change of the zero bit include adding, subtracting, and the logical operations, so this is a generally useful routine.

There is sometimes a problem with these skipping instructions because they skip only one line. The action required in the event of 'no skip' may take two lines or sometimes more. In the example, the finish routine obviously takes several lines, so we had to use a goto instruction to branch to a part of the program where we can have as many lines as we need.

It sometimes happens that we want to put one of two different literal values in the working register, depending on the condition of a bit. For example, a register is to be set to 90h if the bit is clear or to 60h if the bit is set. The problem is that putting a literal value into a register takes two lines, not just one. This is because the value is first put in w (movlw), then moved from the w to the register (movwf).

A way round this problem is to pre-load w before reading the bit:

```
movlw 90h          ; Put 90h in w, ready to spin left.
btfsc randval, 0   ; Test bit <0> of the randval register.
movlw 60h          ; Change setting to spin right.
movwf portb        ; Spin left or right.
```

This example comes from Project 6.4, the mode 4 listing (p. 296). A value is to be sent to Port B to control the direction in which the *Quester* turns. We set w to one of the possible settings (spin left) *before* we test the random bit. Then we test the bit. If it is set we change the value in w to spin right. This needs only the single skippable line. If the bit is set, we skip over the change instruction, leaving the value for spin left as it is. The required setting is now in w, and is moved to Port B to switch on the motors.

The bit test instructions are used in logical operations and when there is input to the program from pressing a button or setting a switch. There are two techniques for using buttons and switches: (1) test the bit in passing, and (2) wait for the bit to take a particular value, either 0 or 1. The Mode Select routine is an example of type (1). The switch has already been open or closed before the program reads its state. The test and the resulting branch are taken care of by the single btfsc or btfss instruction. The Mode Select routine is run at the very start of a program. Why waste the switches on a single routine? The same switches can be used later to select other options. For example, in a game a switch can be turned on to tell the robot 'I have made my move — your move now'.

Waiting for a particular input means putting the PIC into a loop. The simplest way to do it is this:

```
waiting  btfsc portb, 4     ; Test input from switch at RB4.
         goto waiting       ; Test again if input is 1.
... continue with program
```

This routine waits until the input goes low. If there is a weak pull-up on the input (p. 139) we need only place a switch between the pin and the 0 V line. The program waits for us to close the switch. The same for a push-button.

The simple routine above has its snags, the main one being that PICs run very fast. If we are using a button, and there are several wait routines in the program, the PIC may have run on to the next routine while we are still pressing the button for the first routine. If the programming makes it possible for this to happen, we need to check that the button has been pressed *and released* before the program continues. Expand the routine to:

```
waitinglow   btfsc portb, 4      ; Test input from switch at RB4.
             goto waitinglow     ; Test again if input is 1.
waitinghigh  btfss portb, 4       ; Test input from switch.
             goto waitinghigh   ; Test again if input is 0.
... continue with program
```

These routines make the processor wait in a loop. It is not able to do anything else while waiting. Sometimes this is the way we want it. At other times we would like it to do something else while it is waiting. For example, it might flash an LED until we press the button to stop it doing that and do something else instead. In this case we need a routine to sample the button setting at frequent intervals.

Polling is one way of doing it. Keep on sending the PIC to test the button input at such frequent intervals that pressing the button appears to have an immediate effect. Here is a possible solution:

```
flash  bsf portc, 0       ; Turn on LED connected to RC0.
       call delay         ; On for 0.2 s.
       bcf portc, 0       ; Turn LED off.
       call delay         ; Off for 0.2 s.
       btfsc portb, 4     ; Test input from button at RB4.
       goto flash         ; Button not pressed, do another flash.
   ... continue with program
```

Polling is the easiest technique to follow. You know exactly when the input is going to be read. In this case you know that the flashing loop will end with the LED off.

The alternative technique is to use an interrupt, using interrupt-on-change. The problem is that you can not be certain at what stage in the flashing loop this will be triggered. You need to program the PIC to jump out of the loop and switch off the LED before you continue with the program. Also there is the matter of testing the input channels to find out which one caused the interrupt. Interrupts are a great feature in some programs, but are probably better avoided if simple polling will do the job almost as well.

Steering a mobile robot

There are two main techniques for steering a three-wheeled robot:

- Two drive wheels on the same axle (or perhaps with a differential gear) turned by a single motor, and usually at the rear; a single free-running steering wheel, usually at the front, with a motor to turn it in the required direction. The arrangement is that of a child's tricycle. It is used in the *Android*.

- Two drive wheels on separate axles, each with its own motor. The wheels are usually about half-way along the chassis, to give the robot the ability to spin on its centre. The third wheel is a castor. It is similar to steering a tracked vehicle, such as a tank. The programming for tracks is the same as for wheels. This technique is used in the *Quester*.

There is a third technique, used in the *Scooter*, that has a castor which automatically steers the vehicle to one side when it runs in reverse. This extremely simple method is easiest to build and does not need programming, but has limitations.

Two-wheel steering relies on switching the motors individually into forward or reverse. The table opposite lists five possible actions. These are controllable by digital output: a motor is either running or stopped. If it is running, it is either running forward or it is running in reverse. We could program it for analogue control, in which the motors are made to run at different speeds. Then the robot could follow a curved path. But analogue-based programming is complicated and digital is generally good enough.

The technique is to execute the intended curved path as a series of short straight runs alternating with frequent but small spins in the required direction.

Left motor	Right motor	Result
Stop	Stop	Stop
Forward	Forward	Forward
Forward	Reverse	Spin right
Reverse	Forward	Spin left
Reverse	Reverse	Reverse

Digital steering.

To program a path curving to the left, for example, the actual path is a series of forward runs lasting about 0.2 s each (turn on both motors, call *delay*). Between each straight segment there is a short spin to the left. Breaking up the path like this has another purpose. It gives the opportunity between segments to poll the sensors, to check that the robot is heading in the right direction.

Mode 3 of the *Quester* programs is an example of this. The robot is programmed to follow a curved black line. The flowchart on p. 291 shows that in between the short forward-moving segments it checks its sensors to make sure it is still on the line. If it is not, it corrects the error by short spins to the left or right.

The motors are controlled by four digital outputs from Port B. The table on p. 277 gives the details. The left motor is controlled by two outputs, A (from RB7) and B (from RB6). These go to the motor power control board which works as described in Part 4. To run the left motor in a forward direction we make A high and B low. For the reverse direction A is low and B is high. Putting these settings into bytes to send to Port B, we need 10000000 for forward and 01000000 for reverse. In hex these are 80h and 40h respectively.

To make the robot run forward we need to switch on the right motor too. This is controlled by two outputs from Port B, C (from RB5) and D (from RB4).

The complete code byte for running both motors forward must make both RB7 and RB5 high. The code is 10100000, or a0 in hex. To put both motors in reverse, make RB7 and RB5 low, and make RB6 and RB4 high. The code is 01010000, or 50 in hex. Code 00h makes all control inputs low, stopping both motors.

Typical instructions for setting both motors into forward drive are:

```
movlw a0h          ; Code to w register.
movwf portb        ; Both motors turned on.
call delay         ; 0.2 s.
clrf portb         ; Stop both motors.
```

This gives a 0.2 s segment, but we could call it *longdelay* instead, putting a suitable value into w before calling. Or we could let the robot run until a given event interrupts it.

This way of programming depends on the fact that Port B has only four channels (which is why the four LSBs of the code are always low). In another robot you may be controlling the motors from Port A or Port C and want to use the other channels as outputs for other purposes. As an example, suppose the motors are controlled by bit <7:4> of Port C. The same codes apply, and sending a0h to Port C will turn the motors on, but it will turn off all devices connected to the other four outputs. The easiest way is to set or clear the bits individually, using bsf and bcf.

Programming motors switched by relays (pp. 96-97) is similar, but the codes are different. As in transistor switching, each motor is controlled by two bits, but one bit controls on/off and the other controls direction. For a single motor the control outputs are:

On/off relay (1 = on)	Forward/ reverse relay (1 = fwd)	Code (<1:0>)	Result
1	1	3h	Forward
1	0	2h	Reverse
0	0 or 1	0h or 1h	Stop

Motor control with relays.

Steering with two motors is done by doubling the code digits. So forward is 0fh, reverse is 0ah, spin left is 0ch, spin right is 0eh, and stop is 00h.

In Project 6.5 we use three motors switched by relays to move the x-frame, the y-frame and the tool to different places in the working area. These winch motors are switched on one at a time and wind out or wind in the winch cord. For a single motor, the control settings are:

In/out relay (1 = wind out)	On/off relay (1 = on)	Code (<1:0>)	Result
0 or 1	0	0h or 2h	Stop
0	1	1h	Wind in
1	1	3h	Wind out

Winch motor control with relays.

The codes in the table apply when the relays are controlled by bits <1:0>. In the *Gantry* project the wind in/out is controlled by RC7 and the three motors by RC4 (M1, y-winch), RC5 (M2, x-winch) and RC6 (M3, tool winch). The corresponding codes are as listed on p. 339.

Controlling stepper motors is described on pp. 98-100.

Detecting objects

The ability to detect an object that is not in physical contact is often an essential element of a robot's behaviour. A light sensor such as an LDR is used for detecting a source of light at a distance, and an ultrasonic sensor can detect relatively large solid objects up to several metres away. In this section we look at a way of detecting a small object that is a few centimetres away. It might be a game piece that is to be picked up by a gripper.

To sense whether or not there is an object ready to be picked up we employ a simple strategy. There is an LED with its beam directed at where the object is expected to be. The object reflects light from the LED back toward the robot. There is an LDR to sense the reflected light. There may also be other sources of light in the room and we need to be able to distinguish between light coming from these sources and light reflected from the object.

The solution is to flash the LED alternately on and off and to sample the output from the LDR when the LED is on and when it is off. The diagram below shows what happens. The LDR is driven from an ouput channel which is alternately set to 0 and to 1 to flash the LED on and off. The flashing routine should include short delays to give the LDR time to respond. The LDR is connected to an analogue input channel sampled by a comparator. Instead we could use a digital input from an op amp IC, as on p. 73. In either case, output from the comparator falls when the LDR receives light reflected from the object or from other sources.

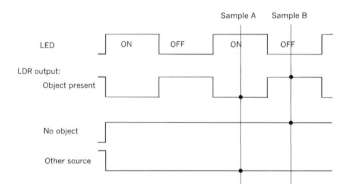

The output from the comparator under different conditions.

If an object is present, the signal from the comparator falls when the LED is on, and is 1 when the LED is off. If no object is present, the comparator output remains constant at 1. This happens too if the object is too far away to reflect enough light back to the LDR. A source of bright light causes a continuous low output from the comparator, remaining low when the LED is off. These three cases are identified by the routine illustrated in the flowchart opposite.

The flowchart sets out the simplest routine. The result of this routine is to be placed in a register labelled `object`, which acts as a flag. The value in the register is to be 01h if an object is found.

The comparator is programmed to give a high output (1) when the LDR is receiving light. The LED is switched on and, after a short delay to give the LDR time to respond, we test the output from the comparator, as described on pp. 187-189. If the output is high, the LDR is receiving light, either by reflection from an object of from another source. Following the 'Yes' branch, we sort out these two possibilities by switching off the LED and testing again. Now we expect a low output from the comparator. In this event the `object` variable is set to 01h, indicating that an object has been detected.

The program would probably continue with a routine to pick up the object if `object` is set to 01h, or to do something else if it is not.

If either of the tests gets a 'No' response, the routine runs on to the end, where `object` is cleared to 00h. Note that we must actually clear it. It might have been set to 01h as a result of a previous test.

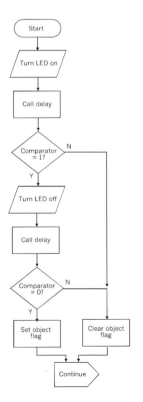

Avoiding objects

Before an object can be avoided it has to be detected. This operation is essentially the same as the object detection routine, but, instead of going on to pick up the object, the robot backs up and steers away from it. A practical example of this is described and listed for the *Scooter* on pp, 194-200.

Bumpers, as used in the *Quester,* are the alternative sensors for object avoidance. Programming them is easy as all that has to be done is read a one-bit input and take appropriate action.

Right: Detecting an object. It could happen that there is another source of light that stops shining on the LDR at the same instant that the LED is switched off. Perhaps this is unlikely, but if you want to guard against this error, repeat the two tests before setting or clearing the flag. A 'No' result from any one of the four tests clears the flag.

Musical tones

If your robot has a speaker, it can make music, or at least can produce a beep or two of given pitch and length. This can be a way of getting the robot to talk to you.

The *tonesound* routine operates by switching the speaker on and off at a given frequency for a given number of times. The routine needs three values to be stored in registers before the routine is called:

- `data1` and `datafinal` set the frequency of the tone. `Data1` is the outer loop variable for two nested loops, the inner loop variable is set to 0Fh. These loops are followed by a final loop which acts as a fine adjustment of the frequency, and has `datafinal` as its variable.

- `lendata` sets the length of the tone. This is the loop variable of the inner one of two nested loops, used in the half (= half period) subroutine. The outer loop counter is set to 05h.

Assuming that the oscillator frequency of the PIC is 4 MHz, calculate the half-period in microseconds:

$$half = 500\ 000 / frequency$$

By trial and error find a pair of values for `data1` and `datafinal` for which:

$$(50 \times data1) + (3 \times datafinal) + 15 = half$$

For a tone of suitable length, calculate:

$$lendata = frequency \times 0.1$$

You can alter the length of the tone by varying `lendata`, but the equation above gives a good starting point.

As an example, here is the calculation for a 1 kHz tone:

$$half = 500000 / 1000 = 500\ \mu s$$

A few trials gave us

$$(50 \times 9) + (3 \times 12) + 15 = 501$$

This is as close as we can get, so data1 = 9 and datafinal = 12.

A suitable value for `lendata` is 1000 × 0.1 = 100.

The listing opposite and on p. 152 uses this data to produce a 1 kHz tone.

```
; Registers needed for playing a tone,

        outloop      equ 22h
        lendata      equ 23h
        inloop       equ 24h
        loop1        equ 25h
        loop2        equ 26h
        data1        equ 27h
        datafinal    equ 28h
        finalloop    equ 29h

; Setting up the registers

movlw 064h                  ; lendata = 100 decimal.
movwf lendata
movlw 09h                   ; data 1 = 9.
movwf data1
movlw 0ch                   ; datafinal = 12 decimal.
movwf datafinal


playit
    movlw 05h               ; Set outer loop counter.
    movwf outloop

next1
    movf lendata, w         ; Set inner loop counter.
    movwf inloop

next2
    bsf portc, 0            ; Speaker on.
    call half
    bcf portc, 0           ; Speaker off.
    call half
    decfsz lendata, f       ; Count down inner loop.
    goto next2
    decfsz outloop, f       ; Count down outer loop.
    goto next1
    return
```

Playing a tone (continued overleaf).

```
half
     movf data1, w
     movwf loop1              ; Set outer counter.
next3
     movlw 0fh
     movwf loop2              ; Set inner counter.
next4
     decfsz loop2, f          ; Count down inner loop.
     goto next4
     nop
     decfsz loop1, f          ; Count down outerloop.
     goto next3
     nop
     movf datafinal, w        ; Set final loop counter.
     movwf finalloop

next5
     decfsz finalloop, f      ; Count down final loop.
     goto next5
     nop
     return

     end
```

This continuation of the listing on p. 151 is the half period subroutine

Begin the program by setting Port C <0> as an output. A circuit for driving a speaker is shown on p. 90.

This listing can be made a subroutine of a more complex routine that puts a set of values in the data registers and calls `playit`. In this way you can make your robot play tunes. The table opposite lists data values for obtaining some of the notes on the musical scale. By programming a sequence of such tones you can invent a tone 'language' that the robot uses to express its 'feelings'. A longer sequence of tones becomes a song.

On the opposite page there are flowcharts of the two subroutines. These show the various loops involved in producing a tone of a given pitch and duration. To play more than, say, four tones in succession it is best to have the values stored in a look-up table. This is described in the next section. A practical example of a tone playing routine is given for the *Android* (pp. 230-232).

Note	data1	datafinal	lendata
C	024h	020h	02Ah
C'	010h	02Fh	054h
E'	0Ah	051h	06Ah
G'	0Ah	029h	07Eh
C"	08h	015h	0A8h
G"	06h	01h	0F9h

Variable values for musical tones, to be loaded into the three registers before calling playit.

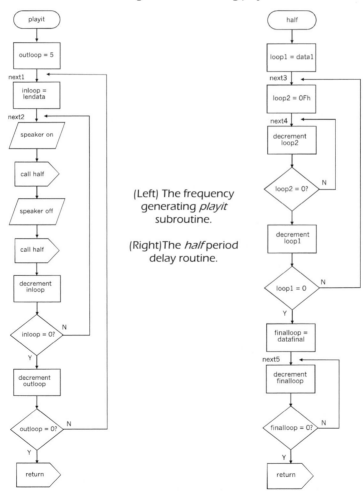

(Left) The frequency generating *playit* subroutine.

(Right) The *half* period delay routine.

Look-up tables

When a routine needs to access data in a systematic way, the best approach is to put the data in a look-up table. The table is a block of data in a subroutine, or a block of EEPROM memory. It might contain a set of numeric values, for example the variables used in the *playit* subroutine. It might contain a set of codes for different arrangements of pieces on a game board. It might contain the coded results of attempts to learn a maze. Stored in a look-up table, the data is quickly accessed.

Here, we look at how to put the data in a subroutine. To access the data we set up a variable which tells the PIC which particular item of data is to be read. Because it points to an item, this variable is named pointer. Pointer is stored in a General Purpose register and given an initial value of zero:

```
pointer   equ 30h        ; Any other unused address is OK.
clrf pointer             ; Reset to 0.
```

The subroutine is best placed early in the listing, particularly if there is a lot of data. This avoids the risk of the table spilling over into two blocks of program memory. The first line is:

```
addwf pcl, w
```

Before the subroutine is called the value of pointer is placed in the working register. To read the first byte, the pointer is 00h. To read the second value it is 01h and so on. On the first line of the subroutine, the program counter is therefore increased by the value of the pointer. Normally the PIC would continue automatically incrementing the program counter and go on to the second line of the subroutine. If the pointer is 00h, this is what happens, but if the pointer is more than zero, the PIC jumps further. It jumps to the line to which it is pointed.

All of the lines of the subroutine, apart from the first, are of the form:

```
retlw, 0ah
```

The line loads a value (in this example 0ah) into the working register and sends the processor back to the main program. There the value in w is used in the program that called the subroutine.

As an example here is a look-up table for the codes used in a 7-segment display.

```
sevenseg
        addwf pcl, f          ;  Jump to the required line.
        retlw 03fh            ; Code for '0'.
        retlw 006h            ; Code for '1'.
        retlw 05bh            ; Code for '2.
        retlw 04fh            ; Code for '3'.
        retlw 066h            ; Code for '4'.
        retlw 06dh            ; Code for '5'.
        retlw 07dh            ; Code for '6'.
        retlw 003h                ; Code for '7'.
        retlw 07fh            ; Code for '8'.
        retlw 06fh            ; Code for '9'.
```

A look-up table of codes for a 7-segment LED or LCD display. The first bit of each code is 0 and the other seven bits set the segments to be on (= 1) or off (= 0). From left to right the bits control segments gfedcba.

This subroutine is intended for a seven-segment display driven through an 8-bit port, such as Port C. To use the 7-segment we load w with the number that is to be displayed and call *sevenseg*. The processor returns with the corresponding code in w. This is loaded into Port C and the numeral is displayed.

A practical example of using a look-up table appears in the tune playing program of the *Android*.

Using two processors

In a system based on distributed processing, two or more controllers share the processing tasks. One of the benefits of this arrangement is to increase computing power and speed. It may also be a way of increasing the number of inputs and outputs of the system to cope with a large number of motors or sensors, for example. Programming this type of system is complicated because the controllers are operating simultaneously and need to be synchronised with each other by a set of hand-shaking signals. A rather simpler system is illustrated in the drawing overleaf.

This is based on the idea of the two PICs, in separate locations but linked by wire or radio, which operate alternately. One waits or performs minor tasks while the other is busy with a more complex task. That completed, the formerly busy PIC signals to the waiting PIC, which gets busy while the other rests. A system such as this (see the diagram) is easier to program because only one processor is active at any given time.

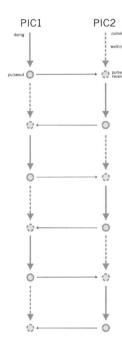

Programming two PICs to operate alternately. A PIC sends a pulse to the other when it has completed its current task. Then it waits (dashed lines) to receive a pulse before resuming its activities.

There are two subroutines for sending and waiting to receive signal pulses:

```
pulseout
    bsf portc, 0
    call delay
    bcf portc, 0
    return

pulsein
    btfss portc, 1
    goto pulsein
    return
```

The controller sends the pulse from channel RC0, and receives pulses through channel RC1. This style of programming is easy to plan but it depends on both controllers being programmed to follow a predictable sequence of actions.

A single pulse has limited meaning. Most times it means 'My job is done — now its your turn'. Sending a coded byte gives the receiving controller far more information. A single byte can take any of the values 0 to 255, so it can have up to 256 different meanings. For instance, if the *Android* has an extra motor to move its left arm, the control of the motors could be given to a Slave PIC and the rest of the processing is done by the Master PIC. The Master sends the Slave signals, telling it which motor to switch on, in which direction to run it, and possibly at what speed. The Master sends one of a pre-defined set of coded bytes. The Slave takes the appropriate action.

The USART is used to transmit the byte serially from Master to Slave. The RB7 channel (pin 10) on the Master is connected to the RB5 channel (pin 12) on the Slave. An ordinary wired connection will do, provided that it is not too long and that the two controllers are operating on the same power supply. It may be advisable to use a light duty screened cable for the connection if the distance is more than a few centimetres. Connect the screen to 0 V at *one* end.

The listing of the Master program includes the segments for configuring the USART for enabling it, and for transmitting a byte. The byte must be in the working register before calling the *transmit* subroutine.

The listing of the Slave program includes the segments for configuring the USART, for enabling it, and for receiving a byte. If a new byte has been received, it is in the RCREG register and the zero flag of STATUS is set to '1'.

Mathematical operations

The instruction set of the PIC includes the two essential operations, addition and subtraction. These can be used to perform other operations, such as multiplication and division. Humans are familiar with the decimal system, so their algorithms for multiplication and decision are based on this. Logic systems are based on the binary system, so it makes sense to use binary instead.

Multiplication by 2 is simply a matter of shifting the binary digits to the left. For example, binary '0001 0011' equals 19 decimal. When shifted one place to the left it becomes '00100110'. The LSD (least significant digit) is filled with a zero. The value is now 38 decimal, which is double the original value.

Shifting digits to the left is performed by the `rlf` instruction. The LSB is filled with the carry bit (bit <0> of STATUS). When using `rlf` for multiplying, clear the carry bit first.

The MSD (most significant digit) is shifted into the carry bit. In the example on the previous page this was a '0', so nothing was lost in the rotation. If we know that this bit can never be a '1' multiplying by 2 is simply a matter of clearing carry and rotating left once. If the MSB is a '1', it must be accounted for by moving it into a register that holds the upper byte of a 2-byte variable.

The 2-byte variable could be stored in two registers `valh` and `vall`. Begin by clearing both, and also carry. Then move the value to be multiplied into `vall`.

```
      clrf  valh              ;  Clear the registers and carry.
      clrf  vall
double
      bcf   status, 0         ; Clear carry.
      rlf   vall, f           ; Multiply by 2, with carry.
      rlf   valh, f
      bcf   status, 0         ; Clear carry ready for next operation.
      return
```

Multiplying by other amounts is simple if the multiplier is a power of 2. For instance, to multiply by 8, run the algorithm above three times. Multiplying by amounts that are not a power of 2 is done in stages.

Opposite is a routine for multiplying an amount in a single byte, `vall`, by an amount in a second single byte, `multiplier`. In this example, it multiplies 31h by 12h to obtain 372h.

The bits of the multiplier are taken one at a time, by *right*-rotating them. If the bit is '0', nothing is done except to left-rotate `vall` ready for the next stage. If the bit is '1', the amount in `vall` is added to `templ`. This is a variable in which the value of the product is gradually accumulated.

At each stage in this process, any overflow from `vall` is rotated into carry and from there rotated into `valh`, so the result is a double-bit value, maximum 65535 in decimal. There is no provision for overflowing into a third byte, but a two-byte maximum should be large enough for most applications of this routine.

```
start
  movlw 031h           ; An example.
  movwf vall
  movlw 12h
  movwf multiplier     ; An example.
  movlw 08h            ; 8 bits to a byte.
  movwf bitcount

multiply
  bcf status, 0
  clrf valh
  clrf templ
  clrf temph
times
  rrf multiplier, f  ; Take bits of multiplier, in order.
  btfss status, 0    ; If carry = 1.
  goto iszero        : If carry = 0.
  movf vall, w
  addwf templ, f     ; Add value (low byte)to temp.
  btfsc status, 0
  incf temph, f
  movf valh, w
  addwf temph, f
iszero
  call double
  decfsz bitcount, f
  goto times

finish goto finish

double
  bcf status, 0
  rlf vall, f          ; Multiply, with carry.
  rlf valh, f
  bcf status, 0        ; Clear carry ready for next operation.
  return

end
```

A routine for multiplying two single-byte values to obtain a double-byte result.

Multiplication can also be done by repeated addition of the original value, but the method given above is quicker.

Dividing by two and by multiples of two is easily done by rotating right and then clearing the carry. Division by other amounts is more easily done by repeated subtraction. The divisor is subtracted from the dividend (the number to be divided) until a negative result occurs. The number of subtractions is counted and this is the quotient (dividend/divisor). The routine overleaf shows a single-byte division routine.

```
start
    movlw 84h              ; Example.
    movwf divid            ; Dividend.
    movlw 018h             ; Example divisor in w.
    clrf count
    call divide

finish    goto finish

divide
    subwf divid, f         ; Subtract divisor from dividend
    btfss status, 0
    goto negative          ; If negative result.
    incf count, f          ; Number of subtractions.
    goto divide
negative
    addwf divid, f         ; Restore dividend to last
                           ; positive value.
    bcf status, 0
    rlf divid, f           ; Double this value.
    bcf status, 0
    subwf divid, f         ; Subtract divisor.
    btfss status, 0
    return                 ; If rounding down.
    incf count, f
    return                 ; If rounding up.

    end
```

A routine for dividing one single-byte value by another and obtaining a single-byte result.

The routine returns a result in count. The result is rounded up or down to the nearest integer.

The greater than (>) and less than (<) operators are easily programmed, using the subwf instruction. Value a is placed in a labelled register (call is vala) and value b is in the working register. The result is obtained by reading the carry flag STATUS <0>, and the zero flag, STATUS <2>.

There are three possibilities:

	Carry bit, C =	Zero bit, Z =
a > b	1	0
a = b	1	1
a < b	0	0

Using `subwf` for comparisons.

For the combined operator >= (greater than or equal to) only the carry bit needs to be read.

Similarly, if the value a is placed in w and we use the instruction `sublw`, we can compare a with a fixed value.

Random numbers

It might be thought that random numbers have no place in robot programming, but there are several situations in which they are needed. In games' programs random numbers can simulate the throwing of a dice, or other chance event. In programs in which the robot learns by experience, its initial pattern of behaviour is often a random one, but it learns to modify this to produce a more effective pattern.

Anothe use of random numbers is to make its actions seem more human. A human often *appears* to be acting randomly, simply because we do not know what is going on in their mind. Random numbers, in moderation can add this quality to robot actions.

Although the term 'random' is used here for convenience, the numbers generated by the routine are totally predictable. It is just that the sequence of digits — 0s or 1s — is so long before it repeats itself that the sequence appears to be a random one. The correct term is 'pseudo-random'.

The routine simulates a hardware (pseudo-)random number generator that uses a shift register.

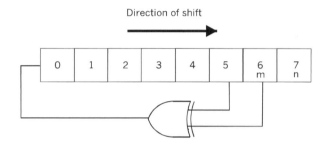

A random number generator can be built from a shift
register IC and an exclusive-OR gate.

The principle is that the contents of two of the registers (m and n) of a shift register are ex-ORed together and fed back into the first register (0). This produces a pseudo-random sequence of digits. The length of the sequence depends on which registers are tapped. As illustrated above and as in the program opposite, with m = 5 and n = 6, the sequence repeats every 127 bits.

The generator has an arbitrary set of bits in its registers to start with, when random (the variable that is equivalent to the shift register) is given the value held in seed. This has at least one '1' in it. If seed is 00000000, the ex-OR gates produce nothing but zeros.

The next step is to assign to bitm and bitn the values at registers 5 and 6. These are ex-ORed together by the xor instruction (the equivalent of the ex-OR gate). The result of this ('0' or '1') is stored in w, and is added to the literal constant ffh. If the result of ex-ORing is '0' the value in w remains ffh and the carry flag stays clear. If it is '1' the value changes to 00h and the carry flag becomes '1'. The rotate left instruction then feeds the carry flag into register 0, producing a new set of values in random.

```
        movlw 0DBh          ;Seed (arbitrarily chosen).
        movwf random

monte
        clrf bitn           ;Clear bit registers.
        clrf bitm
        bcf status, 0       ;Carry = 0.
        btfss random, 5     ;Is n = 1?
        goto findm          ;No: now find m.
        bsf bitn, 0         ;Yes: make bitn = 1.

findm
        btfss random, 6     ;Is m = 1?
        goto xorem          ;No: go to 'logic gate'.
        bsf bitm, 0         ;Yes: make bitm = 1.

xorem
        movf bitn, w
        xorwf bitm, w       ;xor bitn with bitm.
        addlw 0ffh          ;Set carry if w = 1.
        rlf random, f
```

To read a random digit, use btfss or btfsc to obtain the value of one of the digits of random. Then take action according to wheters the bit is '0' or '1'. The bits can also be sampled in groups of two or more to obtain random numbers 0 to 3 (sampling two bits), 0 to 7 (three bits) and so on.

In the listing the generator always begins with the same seed. This produces the same sequence every time it is run. It is better to have a seed that is obtained by chance. One way of doing this is to start with a known seed but keep the generator running in a loop, waiting for a button to be pressed to stop the generator and continue with the program. Exactly when the button is going to be pressed is not precisely known, so the generation of random numbers begins from an unknown seed.

Calibrating the system

The values of some of the variables used in our programs are valid only for our versions of the robot. This is a problem with motor control. When a drive motor is switched on for a particular length of time (say, 0.2 s), the distance travelled by the robot depends on: the type of motor, the reduction ratio of the gearbox, the supply voltage, the diameter of the wheel and tyre. Your robot will differ from ours in at least one of these respects. The same applies to other variables such as the output from analogue sensors.

Often we can decide how to adjust the variable values by watching the robot in action. A robot running on a straight path may be programmed to run for 10 s by calling the *longdelay* subroutine, telling it to call *delay* 50 times. The working register is loaded with 32h before calling *longdelay*. If, for example, the robot moves twice as far as it should do, amend the listing to load w with half the value, 19h.

Sometimes we need to sample a variable while the robot is in action. For instance, the output of an A-to-D converter can be read and its value stored in EEPROM. At the end of the run plug the PIC into the programming board and read out the stored value. Amend the program accordingly.

Software replaces hardware

Often it is possible to use simpler or fewer sensors if we compensate by more elaborate programming and more complex behaviour. Take as an example, the line following behaviour of the *Quester*, which relies on having two sensors.

It would save space, input channels, and cost to have only one sensor. With *one* light sensor, program the robot to keep the sensor directly above the track. If the sensor detects that the robot has deviated from the track, it starts to waggle from side to side, trying to find the track again and stay on it. Program the robot so that it remembers which waggle direction successfully brings it back on track. Then, when the robot next deviates from the track it tries the successful direction first. If it is still on the same curved part of the track, which is likely, this behaviour will help it get back on the track more quickly.

This is just one example of the interdependence of the mechanical, electronic, and programming aspects of robotics. Although they are described in separate Parts of the book, effective design depends on all three, considered together.

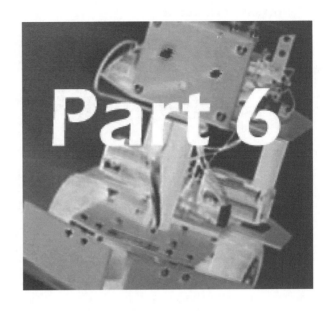

Projects

That was the cookbook — now for some menus ...

Part 6 takes the mechanical, electronic and programming ideas of Parts 3 to 5 and puts them together to make robots. If the individual ideas are the recipes for the courses, the robots are the tasty meals. Part 6 shows five ways of combining some of the ideas, but these are not the only ways of putting them together. For instance, take the infrared sensors of the *Quester* and put them on the *Android*. Then add the maze-solving routines of the *Gantry*. That done, the *Android* will be able to run a maze and learn to solve it. This is just one way of recombining the ideas — there are dozens more.

6.1 The Scooter

This robot is to have low-cost construction and its electronics circuits are to be as simple as possible.

Its modular electronic design makes it a low-cost project in another way. You can change the sensors and re-program the PIC, so that you can in effect build several different scooters for little more than the cost of one.

The panel on the right lists the *Scooter*'s main features, but you can add other features at any time after you have built the essential structures.

Specification

Chassis based on ready-made plastic box.
Runs on 4.5 V or 4.8 V battery.
Three wheels: 2 rear drive wheels, 1 biased castor at front for simple steering.
PIC 16F690A.
Sensors: switch, adaptable light sensors.
Actuators: motor, LEDs, buzzer/siren.

Programming

Hello World!
Calibrating the comparator
Light seeking
Proximity light sensor
Avoiding obstacles
Using an AD converter

Mechanics

Building the robot in a ready-made box gets it up and running that much sooner. It is also a plus feature for those who are not too expert at — or not too interested by — building things. The prototype was assembled in a plastic food storage box with snap-on lid, bought at the local supermarket. The box is about 120 mm square and 50 mm deep. It is made of green transparent plastic and the lid is orange, so it has a striking appearance as it dashes erratically about the room, flashing its LEDs.

Hobby electronics stores are another source of plastic boxes of various shapes and sizes. Choose a squarish one, possibly slightly bigger than the one we used.

When choosing a box, avoid those made of brittle plastic, such as high-impact polystyrene. Food storage boxes are usually made from a fairly rigid yet slightly flexible plastic that is easy to drill and to cut with a craft knife. This type is suitable for the project. Sometimes they soften with the heat of drilling, which leaves a rather ragged edge to the hole, but on the whole it is usually no problem to get a tidy finish.

Slightly more expensive, but suitable for drilling and cutting are the Jiffy boxes or other ABS plastic enclosures sold by electronics hobby shops. Most are black plastic but it is possible to get them in other colours, including transparent ones for people who like to see the works. Choose one which has a squarish shape — many are too long and narrow to be suitable.

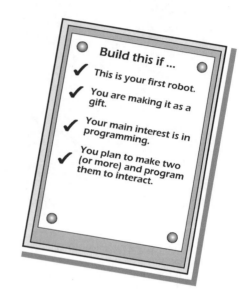

Build this if ...

✓ This is your first robot.

✓ You are making it as a gift.

✓ Your main interest is in programming.

✓ You plan to make two (or more) and program them to interact.

Some have grids marked on the lid, which are helpful when cutting holes. They often have slots for mounting circuit boards, and may have a built-in battery compartment.

Wheels and steering

The *Scooter* has three wheels, which gives it stability on slightly uneven surfaces. It is interesting to contrast the wheel system of the *Scooter* with the very different system of the *Quester*. The *Scooter's* two front wheels are on the same shaft so it needs only one drive motor and only one motor control circuit. This makes the robot easier and cheaper to build.

The third wheel is a castor. How this is constructed depends on what type of small wheel is available. We used a pair of Lego® wheels, with tyres, 25 mm diameter. The wheels clip on to two extensions on a 16 mm square brick. A second brick of the same type is clipped on above this so that the assembly can be bolted to a strip of expanded PVC board (or a strip of brass). The other end of this strip pivots on a bolt that projects from the lid.

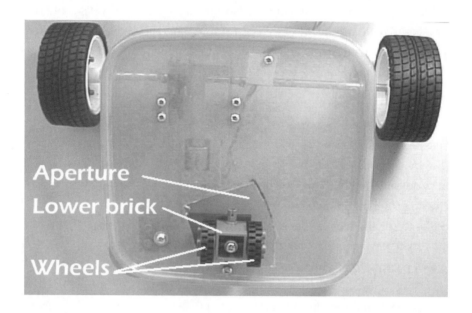

The Scooter's wheel assembly is on the lid, seen here from below (the outer side). It runs on three wheels, in effect, but the rear castor wheel actually is a pair of wheels. The castor is pivoted at a point to one side of the fore-aft centre line. In this photo the castor is in the position it takes up when the robot is running straight ahead.

The robot is steered by an extremely simple mechanism. When moving forward, the castor lever turns to the position shown in the photo above. The *Scooter* always travels straight ahead. To turn, it must reverse a short distance, which makes the castor lever turn to an angled position, as in the photo opposite and in the drawing on p. 179. As a result, the robot backs and turns at the same time.

If it continues to back several times, this action takes it round in a small circle about 400 mm in diameter. Normally it turns through only a small angle before the robot moves forward again, but in a different direction.

This technique is not true steering ability but it is simple to build, works well and is easy to program. You may decide to adopt two-motor steering instead. If so, refer to the description of the *Quester*. Its wheel and motor system can readily be fitted into a plastic box.

Building the wheel system

We chose a typical food storage box, with a snap-on lid. The corners are rounded (see photo). The best way to use this type of box is upside down. The lid (now underneath) has the motor, the drive wheels and the castor mounted on it. The circuit boards and parts of the robot are housed in the box.

An electric motor spins too fast to drive the wheels directly. Use a motor with a built-in gearbox. We chose a 6 V (nominal) DC motor which included a kit of plastic gear wheels. These are assembled to give two different reduction ratios, 1:60 and 1:288. We made up the 1:60 gearing as the *Scooter* is intended to be a fast mover. The axis of the motor is at right-angles to the drive shaft. The drive shaft projects from both sides of the gearbox and each end is coupled to a wheel.

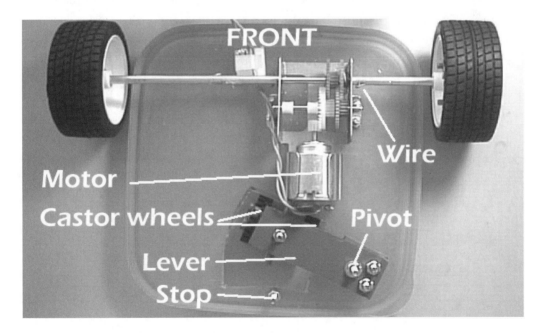

A top (inside) view of the wheel assembly on the lid. The castor wheels are mounted inside the lid but project through a fan-shaped aperture. The lever on which the wheels are mounted turns on a pivot; a long bolt fixed to the inside of the lid. In this photo the lever is in the position it takes when the robot is reversing. The stop peg limits the lever from turning further when in the 'ahead' position (photo opposite). There is no need for a stop peg for the 'reverse' position because the casing of the motor prevents the lever from turning further,

A recurring problem for robot builders is that the drive shaft of the chosen motor and and the hubs of the chosen wheels are not the same diameter. Yet it is essential for the wheels to fit firmly on to the shaft, without slipping. Now may be the time to improvise.

We chose a pair of wheels with a sporty look. They are sold by Tamiya as a 56 mm diameter *Sports Tire* set. The set includes hubs for attaching the wheels to the shaft, but does not include the shaft. To give a stable base, the drive wheels need to be about 170 mm apart, but the output shaft of the gearbox is not long enough. Aluminium tubing, 4 mm diameter was used to extend the shaft at both ends. This is a push-in fit into the hubs and also a push-fit on to the output shaft. The tubing must be long enough to hold the wheel well away from the side of the box.

The 4 mm tubing fits firmly to the output shaft and does not need any support. The tubing did not slip as the drive shaft rotated, but friction grip has a habit of gradually working loose. To prevent this, push the tubing on to the shaft and drill a 1 mm (or smaller) hole through the tubing and shaft. Push a short length of connecting wire through the hole and bend its ends to prevent it from dropping out.

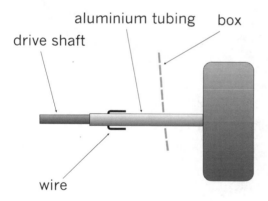

aluminium tubing   box

drive shaft

wire

Coupling one of the drive wheels to the output drive shaft of the gearbox, The wire, threaded through a hole bored in the drive shaft and tubing, eliminates slipping. (Not to scale)

Consider this alternative. As described above, the robot can go only straight ahead or back and turn right. The main pair of wheels could be mounted at an angle so that the robot continuously veers to the left as it goes forward. To go ahead it continually corrects this bias by backing a short way and then continuing forward. To go right it backs a greater distance before continuing. In effect this gives it left-right steering, essential when trying to follow a wall.

Two slots are cut in the side walls of the box, a little wider than the diameter of the tubing.

There are several ways of building the castor. The essential feature is that the wheel(s) is mounted on a lever that is pivoted to the left or right of the fore-aft centre line of the robot. The axle of the wheel(s) is parallel to the lever.

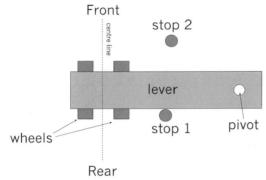

The principle of the eccentric castor.
The view is from above.

The wheel unit is made from two identical wheel-mounting bricks. The wheels rotate on two projections from the sides of the lower brick. The upper brick also has two projections, visible in the photo. These are aligned at right-angles to the projections on the lower brick. A long M3 bolt runs through the central holes in each brick and is used to attach the assembly to the lever.

The wheel assembly is bolted to the lever, which is cut from metal strip or sheet plastic. The wheels are at right-angles to the length of the lever. At the other end the lever pivots on a bolt (see photo overleaf). The length of the bolt and the position of the lever on the bolt are adjusted so that the lid is level when resting on the drive and castor wheels.

The lever may sag under the weight of the robot for two reasons. One reason is that the plastic of the lid is too flexible. In this case, reinforce the lid by bolting to the lid a small square of stiffer plastic at the point where he pivot is mounted.

The prototype *Scooter* did not need this reinforcement. Because the lever has to turn freely on the pivot it can not be bolted firmly to it, so allowing it to sag. It needs to be thicker. The photo shows the solution.

Before drilling the hole for the pivot, two small squares of plastic board are bolted to the lever, using two shorter bolts. Then the pivot hole is drilled through all three layers. The forward stop peg is seen in this photo.

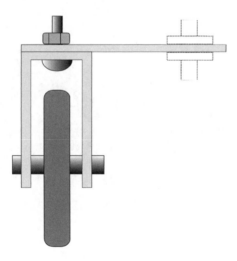

An alternative design for a castor has a single wheel. The mounting could be made from strip aluminium or brass about 10 mm wide.

This completes the mechanical side of the robot apart from mounting the circuit boards when they are ready, and a few other items such as LEDs. Boards are bolted to the lid and the bottom of the box, using 15 mm M3 bolts and nuts.

Summing up:

- Select a box of suitable shape (squarish), size (about 120 mm to 150 mm square), and material (non-brittle, drillable, cuttable plastic).

- Plan the layout of motor, gearbox , drive wheels, and castor. Try to give the robot as much clearance as practicable between its underside and the surface. For stability, the drive wheels should be as far back and as far apart as possible.

- The gearbox ratio should give the *Scooter* a speed in the range 20 cm/s to 50 cm/s.

- Ensure that the drive wheels do not slip and will not come off the drive shaft when driven.

- Cut an aperture in the lid for the castor wheels, and pivot the lever so that it does not sag.

- Position the stop pegs.

- Mount the motor/gearbox/drive-wheel unit.

- Cut the slots in the sides of the box through which the drive shafts pass.

- Finally, check that the robot is level on its three wheels, and that the wheels do not rub against the box.

Shopping list — mechanical

Box, squarish, made of non-brittle plastic, possibly transparent.

DC motor and gearbox with output shaft on both sides.

Pair of wheels, with tyres, about 60 mm diam.

Metal tubing to fit into wheels and fit over drive shaft of motor.

Pulley wheel(s) about 25 mm diam for castor, with shaft about 40 mm long.

Material for making castor wheel bearings.

Nuts and bolts, mainly M3, bolts 6 mm and 10 mm long.

Electronics

The robot is controlled by a PIC16F690 microcontroller. Several other types of PIC could be used with small amendments to the programs. The system is powered by a battery of four lithium metal-hydride AAA or AA cells. These give 4.8 V output (up to 5.2 V when freshly charged) which is within the PICs maximum limit of 5.5 V.

The system is on three circuit boards: the controller board (for the PIC), the motor control board, and the switching board. There are a number of off-board components such as the battery, light-emitting diodes (LEDs) and light dependent resistors (LDRs).

To save space and keep the *Scooter* small and nippy, we used bare-ended, PVC insulated, single-stranded connecting wire. The wire ends are pushed into socket strips on the boards. The wire sold as hook-up wire or bell wire is sometimes a little too thick to go into the sockets. Check this point before buying it.

Controller board

Keeping to the theme of small compact units, the controller board is minimal in design. It is essentially a 20-pin IC socket with socket strips for the connections.

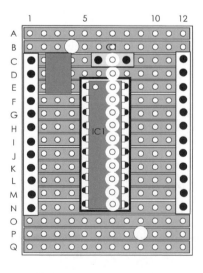

The controller has socket strips on each side for push-in connections. Note the cuts in the copper strips to isolate the pins on the two opposite sides. Sockets at C1, D1 and E1 are connected as a group by running a blob of molten solder between them on the rear of the board. This group is for the positive supply voltage. It is labelled with a patch of red insulating tape. The 0 V supply is at C12, D12, and E12, which are also joined by solder blobs beneath the board. Capacitor C1 is 100 nF miniature polyester.

Motor control board

This has an H-bridge for controlling the direction of the motor. It is mounted on the lid, beside the motor/gearbox unit (photo p. 180).

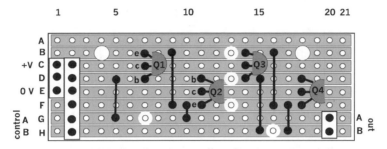

The motor control board has two connecting sockets for the supply lines. This allows it to be daisy-chained to other boards and units in the system.

Connections are made to this board in the same way as to the controller board, by pushing the stripped ends of the wires into sockets. Alternatively, solder the wires directly to terminal pins or to individual sockets that push on to terminal pins. Details of the connections are on p. 178.

Switching board

This is the name given to the board that interfaces the PIC to the sensors and to the actuators other than the motor. There is room for a second board of this kind if the system is expanded later. Schematic and layout diagrams are overleaf.

It is possible to provide current for LEDs directly from a PIC output, but this should not be more than 20 mA. This project uses high intensity LEDs to provide illumination for the light-activated proximity sensors. The LEDs require up to 70 mA each, so must be turned on and off by transistor switches. In the prototype, D1 and D2 are white LEDs, producing 18 cd. The side-facing LED, D3 is a blue one, producing 10 cd.

The buzzer is a semi-mechanical type and takes 40 mA, so needs a transistor switch.

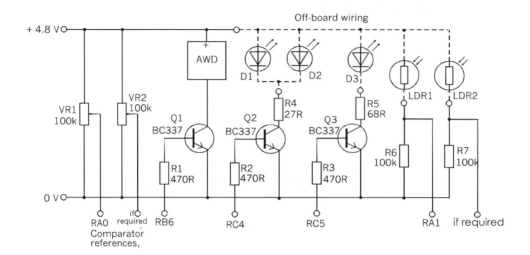

As well as transistor switches for the LEDs and buzzer (AWD), the switching
board has the resistors of the two light sensor circuits, It also has two
optional voltage dividers for use with comparator circuits.

The BC337 transistors specified in the schematic are rated at 800 mA, which is far more
than the recommended LEDs need. Almost any other npn transistor will do (such as a
BC548), unless you are switching several LEDs in parallel and running them at the peak
current of 100 mA or more. The BC548 has the same pinout as the BC337, so the stripboard placing is the same for both types.

The resistance values of R6 and R7 depend on the values of the light-dependent resistors
LDR1 and LDR2. The resistance should be approximately equal to the resistance of the
LDR under the light conditions in which the robot is expected to operate. The output
from the sensor will then vary over a range centred on half the supply voltage.

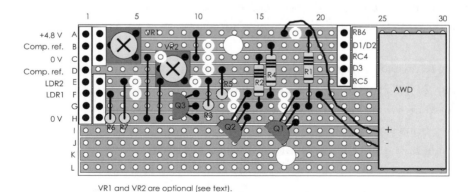

VR1 and VR2 are optional (see text).

Layout of the switching board. The two trimmer resistors, VR1 and VR2, are optional.

The completed Switching Board.

Mounting the boards and off-board items

The three boards are each supported on two M3 bolts. The exact positioning depends on the size and shape of the box. Remember to place the Controller board so that it is easy to remove and replace the PIC when debugging the program. Check that the motor and other items on the lid will not come into contact with the boards when the lid is on the box. Leave enough room for the wiring.

Drill 3 mm holes in the box for the bolts that support the boards. Drill a hole in the bottom of the box for the power switch.

The two headlamp LEDs, D1 and D2, are inserted in 5 mm holes. The holes are in what will eventually be the forward-facing wall of the box. Drill them about half-way up the front of the box, one on the left and one on the right. When drilling these holes angle the drill bit so that the 'beams' of the LEDs will converge to produce a single spot of light on objects about 100 mm ahead of the robot. Drill a single 5 mm hole for D3 on the left side of the robot.

To provide directional sensitivity, the LDRs are inserted into 10 mm pieces of plastic sleeving of suitable diameter. A hole of a size to grip the sleeving is drilled for each LDR. LDR1 faces forward and is mid-way between D1 and D2. LDR2 is directed to the left and is close to D3.

Off-board connections

The boards are connected by single-stranded PVC insulated connecting wires. The insulation is stripped from the ends of each wire for about 5 mm. The table below lists the connections needed. Cut the wires as short as conveniently possible.

| FROM | | Function | TO | |
Board	Socket		Board	Socket
Controller	E1	Positive supply	Switching	A1
Switching	A2	Positive supply	Motor control	C2
Motor control	C1	Positive supply	Power switch S1*	
Power switch S1 (common) *		Positive supply	Battery positive	
Controller	E12	0 V line	Motor control	E2
Motor control	E1	0 V line	Switching	C1
Switching	C2	0 V line	Battery negative	
Controller	L1	Motor control A	Motor control	G2
Controller	M1	Motor control B	Motor control	H2
Motor control	G20	Motor output A	Motor M1 terminal*	
Motor control	H20	Motor output B	Motor M1 terminal*	
Controller	G12	LDR1	Switching	F1
Controller	F12	Comp. 1 ref.	Switching	B1
Controller	N12	Buzzer	Switching	A22
Controller	J1	D1 and D2	Switching	C22
Controller	I1	D3	Switching	E22

In a few cases, indicated by * in the socket columns of the table, the wires are soldered to the terminals.

The positive supply also goes to the off-board LEDs and LDRs. For neatness, this line and the connections returning from the components are best assembled as a unit.

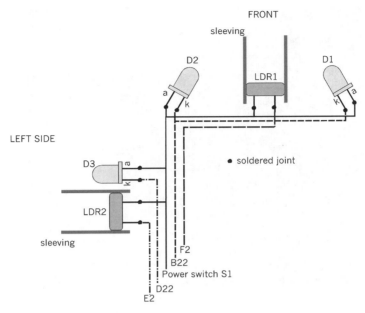

Connections between the Switching board and the LEDs/LDRs
(diagrammatic).

The positive line is drawn in the diagram as a continuous line. It is a length of bare connection wire which runs from the anode of D1 (front right), across the front of the body, to half-way along the left side. Then it continues to the positive terminal of the power switch S1. A small loop is twisted in this line level with each of the components it visits: D1, LDR1, D2, D3 and LDR2. There is insulated sleeving between these points. The leads of the components are cut short, bent into hooks, and hooked into the loops in the positive line. The loops and hooks are squeezed together with pliers and the joints are soldered.

Four other lines run back from the non-positive terminal of each component to a socket on the switching board. One of these serves *two* components, the cathodes of D1 and D2. These connections use insulated single-stranded connecting wire and are wound spirally around the positive line to give the assembly some rigidity.

The wired assembly is supported by two small cable clips, situated on either side of LDR1. These self-adhesive clips do not adhere to the plastic of the box, so they are bolted in place with M2 bolts.

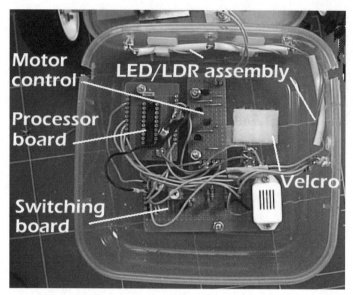

The interior of the box when wiring is finished, as seen from below. Most of the connections are between the Processor board and Switching board, but the wiring has been routed away from the processor socket so that it does not interfere with putting the PIC into its socket and removing it. The connections to the LED/LDR assembly can be seen going off to the left (right in the photo), near the label 'Velcro'. The main switch S1 can just be picked out between the Velcro and the buzzer.

The patch of Velcro is to hold the battery box in place. The box holds four AAA rechargeable NiMH cells and has a matching Velcro patch on its side. It connects to the circuit by a standard PP3 battery connector.

This completes the wiring but it is essential to test everything at this stage. Routine testing is described on pp. 102-103. The main stages are:

- Test for continuity on the 0 V line and positive supply line.

- Test for short circuits between these two lines.

- With the PIC removed from its socket, switch on the power and check that each board and off-board component receives power.

- At the PIC socket apply a positive voltage to each output line using the table opposite to identify test points. The LEDs and buzzer should operate. With a positive and 0 V lead applied to pins 8 and 9, the motor should run forward or in reverse.

- Use a test meter to see that the voltage from the LDRs is a reasonable value with the LDRs exposed to light and shaded.

PIC I/O

For later reference, the table below lists I/O connections to the PIC. A photocopy of this table mounted on card is handy to have on the workbench while assembling and testing the circuit modules.

Port	Name	IC pin	I/O	Connected to	0 =	1 =
A	RA0	19	AnI	Reference voltage (AN0)		
	RA1	18	AnI	Forward light sensor (AN1)		
	RA2	17	I	Program select	Program 1	Program 2
	RA3	4	I			
	RA4	3	I/O			
	RA5	2	I/O			
B	RB4	13	I/O			
	RB5	12	I	Reserved for USART input		
	RB6	11	O	Bleeper	Off	Sound
	RB7	10	O	Reserved for USART output		
C	RC0	16	I/O			
	RC1	15	I/O			
	RC2	14	I/O			
	RC3	7	I/O			
	RC4	6	O	Headlamps	Off	On
	RC5	5	O	Side lamp	Off	On
	RC6	8	O	Motor A	Forward	Reverse
	RC7	9	O	Motor B	Reverse	Forward

The entries are grouped by I/O port and list all the channels available on the PIC16F690. In the fourth column, the I or O indicates whether the channel is to be configured as an input or an output. AnI is an analogue input.

RC0 to RC3 can be used as analogue inputs for the comparators or AD converters if you decide to add more sensors, such as LED2.

The table shows the settings of Motor A and B lines. To stop the motor make both 0 or both 1.

Shopping list — electronic

Controller board:

      C1 polyester capacitor, 100 nf
      IC1 PIC16F690
      20-way d.i.l. turned pin IC socket
      socket strip 2 x 12 sockets
      stripboard 17 strips x 12 holes

Motor control board:

      Q1, Q3 BC639 npn transistor (2 off)
      Q2, Q4 BC640 pnp transistor (2 off)
      socket strip, 2, 3, and 6 sockets
      stripboard 8 strips x 21 holes

Switching board:

      R1-R3 resistors 470R (3 off)
      R4 resistor 27R
      R5 resistor 68R
      R6, R7 resistors 100K (2 off)
      VR1, VR2 horiz. trimmers 100K (2 off), optional
      audible warning device, solid-state
      socket strip, 3, 4, 5, and 8 sockets
      stripboard 12 strips x 30 holes

Off-board components:

      D1 - D3 5 mm light emitting diodes, ultra bright (3 off)
      LDR1m LDR2 light dependent resistors (ORP12 or similar)
      S1 SPST mini toggle switch
      battery holder, 4 x AAA or 4 x AA, with wire or stud terminals
      PP3 type battery connector (if battery box has stud terminals)
      AAA or AA NiMH rechargeable cells (4 off) (preferred)

Miscellaneous:

      Strip of Velcro™ sticky Back®
      Single stranded connecting wire (to fit sockets)
      Solder

Programming in assembler

It is assumed in this section that you are using the PICkit 2 programmer and software and writing in assembler. If you are using another programmer, your listing will be largely the same but with minor differences. If you are programming in PICBASIC, there are versions of these assembler programs on pp. 200-208.

Hello World!

It is almost a tradition for beginners' books on programming to begin with the simplest of all programs — displaying 'Hello World' on the monitor screen. This program is the robotic equivalent. It makes the *Scooter* show off its output capabilities in the simplest possible routine. At the same time it is a way of checking that the output circuits are working properly.

The first thing the *Scooter* does when this program is run, is to stay motionless for about 5 seconds, waiting for everybody to give it their full attention. Then it switches its two headlamp LEDs on for 5 s. Next it runs forward for 2 s, so make sure that you are not in its way. Finally it switches on its side lamp LED for 2 s. The program that produces these actions is listed overleaf.

The file begins with a leader that identifies the program. This is not part of the program. This header is ignored by the assembler because the assembler always ignores anything on a line that comes after a semicolon (;). You can alter this header or leave it out altogether.

The first lines in any code define variables and other initial settings. The listing states the type of controller the program is written for. In this example it is the 16F690. This is followed by the configuration code, which is 0x33c4. Note that the directive __CONFIG begins with *two* underline characters. The listing continues with directives giving the addresses of registers, the code values of 'w' (working register) and 'f' (the current file register), and the labels of the delay subroutines. Depending on the software you are using, you may not need to include all these addresses because they are built into the software or are found by calling on an 'include' file.

```
;*************************************************************
;     Filename:     Scoot01.asm                             *
;                                                           *
;     Hello World!                                          *
;                                                           *
;*************************************************************

        list      p=16F690              ; The controller
        __CONFIG  0x30c4

; Bank0
        status          equ 03h
        portb           equ 06h
        portc           equ 07h
; Bank1
        trisb           equ 06h
        trisc           equ 07h
; Bank2
        ansel           equ 1eh
        anselh          equ 1fh
; Destination code
        f               equ 01h
; Labels
        delay0          equ 20h
        delay1          equ 21h
        delayn          equ 22h

        org 00h
        goto start
        org 04h
        goto start

start
        bcf status, 5       ; Bank0.
        bcf status, 6
        clrf portb
        clrf portc
        bsf status, 5       ; Bank1.
        clrf trisb          ; Port B all outputs.
        clrf trisc          ; Port C all outputs.
        bcf status, 5       ; Bank2.
        bsf status, 6
        clrf ansel          ; Digital input/output.
        clrf anselh         ; Digital input/output.
        bcf status, 6       ; Bank0.
bcf status, 5
```

```
; Program begins here
    clrf portb
    clrf portc
    movlw 019h          ; Delay 5 s. Get ready to watch display.
    call longdelay

    bsf portc, 4        ; Headlamps on.
    movlw 019h          ; Delay 5 s.
    call longdelay
    bcf portc, 4

    bsf portc, 7        ; Go forward.
    movlw 0ah           ; Delay 2 s.
    call longdelay
    clrf portc          ; Stop.

    bsf portc, 5        ; Side lamp on.
    bsf portb, 6        ; Buzzer on.
    movlw 019h          ; Delay 2 s.
    call longdelay
    clrf portc          ; Side lamp off.
    clrf portb          ; Buzzer off.

endit   goto endit

; Subroutines


delay
    decfsz delay0, f
    goto delay
    decfsz delay1, f
    goto delay
    return

longdelay
    movwf delayn
repeat
    call delay
    decfsz delayn, f
    goto repeat
    return

    end
```

Now set up the controller so that it operates correctly with the output circuits (the inputs from the LDRs are not used in this program). Bits <5> and <6> of the STATUS register switch in the banks of Special Function registers. The registers for Port B and Port C in Bank 0, are cleared. Port A is not used in this program. Then switch to Bank 1 to set the tristate registers so that all channels in Ports B and C are outputs.

Unlike some of the earlier PICs, the channels of the 16F690 are analogue by default and have to be set to digital. So visit Bank 2, where ANSEL and ANSELH are located, and clear them both to '0', which makes everything digital.

The actual program begins at the top of p. 185. The port registers are cleared to make certain that the LEDs do not light up immediately. That done, there is a pause, using the *longdelay* subroutine.

The LEDs and motor are switched on or off by setting bits in the port registers. The table on p. 181 lists which bits. Setting bit 4 of Port C, for instance, makes the RC4 output go high and so activates the transistor switch the turns on LED1 and LED2. The 'bit set', instruction (bsf) is used for switching on, and the 'bit clear' instruction (bcf) for switching off. At the end, just to be sure that nothing remains switched on, clear all channels of both ports, using 'clrf'.

Having done its tricks, the *Scooter* is left in a continuing *endit* loop.

To produce the pauses the program calls on the *longdelay* subroutine. This calls on the *delay* subroutine. Both subroutines are at the end of the listing. At the very end of the listing is the essential directive, *end*.

Hello some more

The listing can be adapted and extended to increase the *Scooter* repertoire:
- Make it flash the LEDs two or three times instead of only once.
- Make it give a treble beep before and after it runs forward.
- Make it run backward and then forward in a different direction.
- Make it end the routine by flashing its LEDs in an endless loop.

Seeking the light

In this program, the *Scooter* is set down in a room which has fairly subdued lighting. There is one brighter souce of light which may be either a table-lamp on the floor or (in daytime) a window with a low sill. The robot's task is to locate this source and to move toward it.

The program is also an example of how to use the PIC's comparators. Before going on to the light-seeking program, here is a program intended for running while setting the variable resistor VR1 (p. 176). A diagnostic program such as this is useful for checking that the sensor and voltage reference are working properly.

```
;****************************************************************
;     Filename:     Scoot02.asm                              *
;                                                            *
;     Calibrating the comparator.                           *
;                                                            *
;****************************************************************

        list        p=16F690              ; Define processor

        __CONFIG    0x30c4

; Bank0
        status          equ 03h
        porta           equ 05h
        portb           equ 06h
        portc           equ 07h
        intcon          equ 0bh
; Bank1
        option_reg      equ 01h
        trisa           equ 05h
        trisb           equ 06h
        trisc           equ 07h
; Bank2
        cm1con0         equ 19h
        ansel           equ 1eh
        anselh          equ 1fh
; Destination codes
        w           equ 00h
        f           equ 01h
        z           equ 02h
```

```
; Labels
    delay0        equ 20h
    delay1        equ 21h

    org 00h
    goto start
    org 04h
    goto start

start
    bcf intcon, 7       ; Disable interrupts.
    bcf status, 5       ; Bank0.
    bcf status, 6
    bsf status, 5
    clrf trisb          ; Port B all outputs.
    clrf portc          ; Port C all outputs.
    bcf status, 5       ; Bank2.
    bsf status, 6
    movlw 03h           ; Digital input except RA0 and RA1.
    movwf ansel
    clrf anselh         ; Digital output.
    movlw 080h          ; Enable comparator.
    movwf cm1con0
    bcf status, 6       ; Bank0.
    bcf status, 5

; Program begins here

    clrf porta
    clrf portb
    clrf portc


sample
    bsf status, 6       ; Bank 2.
    movlw 050h          ; Bit 6 = 1.
    andwf cm1con0, w    ; Read output bit.
    bcf status, 6       ; Bank 0.
    btfss status, z     ; Test zero bit.
    goto sample
    bsf portc, 5        ; Side lamp on.
    call delay
    bcf portc, 5        ; Lamp off
    goto sample
```

```
; Subroutine

delay
      decfsz delay0,  f
      goto delay
      decfsz delay1,  f
      goto delay
      return

end
```

After stating the type of PIC and the configuration word, there is a list of labels for special purpose registers. This list is generally not needed if you are running development software that uses 'include' files. There are no interrupt service routines so we go straight to the main program.

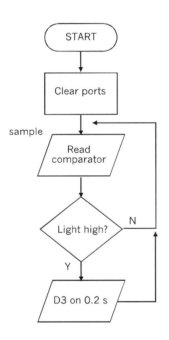

The flow chart shows that the program runs in a continuous loop. Each time around the loop the output of the comparator is read, and the LED is switched on for 0.2 s or not, depending on the reading.

 If the LDR is receiving bright light there is a high input voltage to the comparator. If this voltage is greater than the reference voltage coming from VR1, bit <6> of CM1CON0 is high. The state of this bit is monitored by looking at the zero flag in the status register. If the input voltage is high, bit <6> of CM1CON0 is '0',and the zero flag is set. The instruction 'btfss status, z' results in skippng the next line and turning the LED on for 0.2 s. Then the program loops back to test the bit again. If light is low, bit<6> is '1', the zero flag is clear, and the program loops back without flashing the LED.

Use the program like this. Place the robot so that it is receiving reasonably bright light — as bright as the target lamp to be used when running the light-seeking program. Run the calibration program. Adjust VR1 with a screwdriver so that the LED *just* comes on. The LED should go out immediately if the source of light is removed or reduced.

Now for the light-seeking program. This is based on a simple procedure. The room has low intensity diffuse background light. There is a single bright source of light, such as a table-lamp placed on the floor, or a window with bright sky outside. The *Scooter* is placed on the floor, not facing the light source. The flowchart explains what happens next.

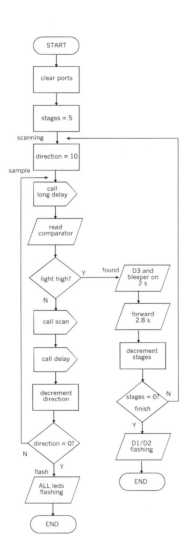

The *Scooter* first tries to find the light source, using the routine beginning at the label 'sample'. It reads the comparator output to determine if the light ahead of it is bright or not bright. By 'bright' we mean that the voltage from the LDR sensor is greater than the reference voltage.

If the scene ahead is not bright, the robot goes into the 'scan' subroutine. It backs a short distance, turning right as it goes, then runs forward a short distance. In effect it has turned a little to the right and is now looking at a different part of the room. For as long as the bright light does not come into its view, it continues to read the comparator output and then scan one step. After 10 such attempts (see bottom of left column) it will have covered a 360° view of the room and failed to find any bright light. It abandons the search and flashes all its LEDs. The program ends at this point.

Left: The light-seeking program gives the robot five attempts to home on a bright light.

If, while it is reading the comparatpr and scanning, the comparator gives a logic high output, the robot knows it is facing toward a bright source. It goes to the 'found' label, where it first of all celebrates by switching on its side LED and sounding its bleeper. Then it rushes forward to get to the source.

However, the angular resolution of its light sensor is rather wide so it may not be heading accurately. After travelling for a few seconds it may be heading to one side of the source. It must aproach the source in stages, checking the direction at the end of each stage. So when it stops moving forward it returns to the 'sample' label and repeats the reading and scanning behaviour.   When it finds the light again it moves forward, perhaps in a new direction.

Altogether it is allowed five 'goes' or stages to get to the lamp. If it does not reach it by then the bottom of the right column of the flowchart shows that it stops and flashes just its headlamp LEDs.

```
;****************************************************************
;     Filename:     Scoot03.asm                                *
;     Scooter seeking light, using comparator.                 *
;                                                              *
;****************************************************************

        list        p=16F690              ; Define processor

        __CONFIG    0x30c4

; Bank0
        status          equ 03h
        porta           equ 05h
        portb           equ 06h
        portc           equ 07h
        intcon          equ 0bh
; Bank1
        trisa           equ 05h
        trisb           equ 06h
        trisc           equ 07h
; Bank2
        cm1con0         equ 19h
        ansel           equ 1eh
        anselh          equ 1fh
```

```
        ; Bits
        w          equ 00h
        f          equ 01h
        z          equ 02h
        ; Labels
        delay0        equ 20h
        delay1        equ 21h
        direction     equ 23h
        stages        equ 24h

        org 00h
        goto start
        org 04h
        goto start

start
        bcf intcon, 7      ; Disable interrupts.
        bcf status, 5      ; Bank0.
        bcf status, 6
        bsf status, 5
        clrf trisb         ; Port B all outputs.
        clrf portc         ; Port C all outputs.
        bcf status, 5      ; Bank2.
        bsf status, 6
        movlw 03h          ; Digital input except RA0 and RA1.
        movwf ansel
        clrf anselh        ; Digital output.
        movlw 080h         ; Enable comparator.
        movwf cm1con0
        bcf status, 6      ; Bank0.
        bcf status, 5

; Program begins here

        clrf porta
        clrf portb
        clrf portc

        movlw 08h
        movwf stages

scanning
        movlw 0ah      ; Direction = 10
        movwf direction
```

```
sample
     movlw 03h
     call longdelay
     bsf status, 6        ; Bank 2.
     movlw 050h           ; Bit 6 = 1.
     andwf cm1con0, w     ; Read output bit.
     bcf status, 6        ; Bank 0.
     btfss status, z      ; Test zero bit.
     goto found
     call scan
     call delay
     decfsz direction     ; Counting down the scans.
     goto sample          ; Try again.
     goto flash
found
     bsf portc, 5         ; D3 on.
     bsf portb, 6         ; Bleeper on.
     movlw 0ah
     call longdelay
     clrf portc           ; D3 off.
     clrf portb           ; Bleeper off.
     bsf portc, 7         ; Forward.
     movlw 0eh
     call longdelay
     clrf portc           ; Stop.
     decfsz stages
     goto scanning
     goto finish
flash
     bsf portc, 5         ; Side lamp on.
     bsf portc, 4         ; Head lamps on.
     movlw 03h
     call longdelay
     clrf portc
     movlw 03h
     call longdelay
     goto flash
finish
     bsf portc, 4         ; Headlamps on.
     movlw 03h
     call longdelay
     clrf portc
     movlw 03h
     call longdelay
     goto finish
```

```
        ; Subroutines

scan
        bsf portc, 6        ; Reverse.
        movlw 05h
        call longdelay
        clrf portc          ; Stop.
        call delay
        bsf portc, 7        ; Forward.
        movlw 05h
        call longdelay
        clrf portc          ; Stop.
        return
```

The program also needs the *delay* and *longdelay* subroutines (p. 185), followed by the 'end' directive.

This program employs an external reference. This allows the triggering level to be reset with the screwdriver instead of having to edit the program and reassemble it.

Robots differ in their motors and wheels so that 10 `directions` might not cover the whole circle. If so, edit the program for more `directions` (first line after the 'scanning' label).

If you are running the *Scooter* in a large room (such as an empty double garage) it could be allowed a few more stages to get to the lamp. Amend the value given to `stages` (lines immediately before the 'scanning;' label).

Avoiding obstacles

For this event, the room needs to have plenty of floor space and to be evenly lit. It can have fairly bright overall lighting as long as it is not stronger than the beams from the two forward-facing LEDs. The program illustrates a technique for light-based proximity detection.

The technique uses the PIC's analogue to digital converter to measure the light received at the forward-facing LDR while the forward-facing LEDs are on. It compares this reading with the reading from the LDR when the LEDs are off.

The LEDs are mounted on the body of the robot and directed on a single central point about 100 mm ahead of the robot. If an object is present at this point, and assuming the object is large enough and fairly reflective, the amount of light received by the LDR will be significantly greater than normal. The robot detects this situation and takes avoiding action.

The robot will respond also when it approaches a wall or furniture, so it is able to scoot about the room indefinitely, though it may eventually get stuck in a corner. You could try to program it to get out of such a trap.

The program listing (pp. 198-199) twice includes a routine for recording the AD output in EEPROM. This was written into the program during development to monitor the data processing. After running the program to test the behaviour of the robot, the PIC is returned to the programmer board, for the stored data to be read. These routines are not a necessary part of the final program and there is no need to type them in. However, they are very helpful at the development stage.

The main program loop is relatively simple:

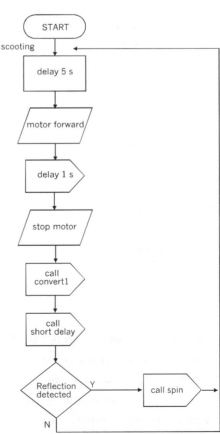

In the main loop of the obstacle avoiding program, the robot moves forward for a short distance, then flashes its LEDs on and off (in the convert1 subroutine). If it detects an obstacle, it backs and turns (in the spin subroutine) but, if it does not, it carries straight on.

The interesting part of this program is the *convert1* subroutine.

The action of this subroutine is to return a value stored in the variable *flags*. If no obstacle is detected ahead the value returned is 1. If not, it is 0.

The subroutine begins by setting *flags* to 0. It then switches on the headlamps (D1 and D2). There is a short delay to allow the AD circuit to equilibrate, and then the PIC goes to a further subroutine (*adread*) to perform the AD conversion. It sets bit<1> of the ADCON0 register to start the conversion. Then it waits in a loop until bit<1> is cleared, indicating that conversion is complete.

Returning to *convert1*, it moves the result of the conversion from its working register to a register labelled *reflect*.

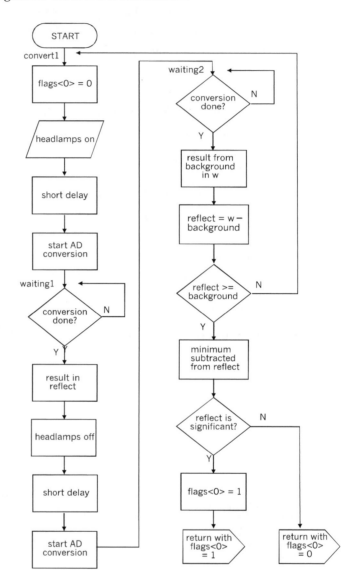

The *convert1* subroutine of the obstacle avoiding program.

The whole procedure is then repeated with the LEDs switched off. This time the PIC returns from *adread* with a value in its working register which represents the level of the background illumination of the room at that time. Based on the two readings, with headlamps on and headlamps off, the PIC decides if there is an obstacle ahead or not. Consult the listing to see the details.

The first step in processing the readings is to subtract the background reading from *reflect*. If this gives a negative result, something odd happened during the reading and the PIC is sent back to try again. The program finds out whether the result of a subtraction is negative by testing the carry bit (bit<0>) of the STATUS register is 0. If the result is positive or zero, the carry bit is 1. So the line 'btfss status, 0' performs the check and sends the controller back to the beginning or to the next stage of processing.

There may be random changes of illumination between the two readings that are not due to reflection from an object, and these should be ignored. We set a minimum value for what is a significant difference, subtract this and test to see if this still leaves a positive value in *reflect*. If so, flags<0>is set to 1. If not, flags<0>> is left as 0.

This completes the processing and the subroutine returns to the main loop with flags<0> set or clear.

The value subtracted from *reflect* controls the sensitivity of the processing. Too big and the robot will fail to detect an object; too little and it will repsond to small changes in ambient light levels that are not the effect of an object ahead. Edit the program if you think the subtracted value is wrong for your robot. This is the purpose of the routines that store the data in EEPROM, for you can read what values were returned from the two readings.

In this program we are not interested in 10-bit precision. This is why the converter was configured to right-justify the result, putting the eight most significant bits in ADRESH. The line 'bsf eecon1, 7' does this. On returning from *adread*, the result is in ADRESH, and the two least significant bits (in ADRES) are ignored.

The listing begins in the usual way by stating the controller type and its configuration. It then lists the equates, but does not need them all. The full list is on p. 135 but can be shortened by deleting unused register names. This program uses:
Bank 0: STATUS, PORTA, PORTB, PORTC, INTCON, ADRESH, ADCON0.
Bank 1: OPTION_REG, TRISA, TRISB, TRISC, WPUA, ADCON1.
Bank 2: EEDAT*, EEADR*, ANSEL, ANSELH.
Bank 3: EECON1*, EECON2*.
Those marked * are needed only if you are wanting to write results in the data memory.

Equates are used for w, and f, and for the variables, *delay0, delay1, delayn, reflect* and *flags*.

The Robot Builder's Cookbook

The instructions for setting up the controller begin at the *start* label. They are the same as those in the listing on pp. 191-192. After this comes the program, listed below.

```
        ; Program begins here

                movlw 019h          ; Delay 5 s.
                call longdelay

        scooting
                call shortdelay
                bcf flags, 0
                bcf portc, 6        ; Motor forward.
                bsf portc, 7
                movlw 05h
                call longdelay
                bcf portc, 7        ; Stop.
                call convert1       ; Read forward sensor.
                call shortdelay
                btfss flags, 0      ; If reflection.
                goto scooting
                call spin
                goto scooting

        flash
                bsf portc, 5        ; Side lamp on.
                call delay
                bcf portc, 5        ; Side lamp off.
                call delay
                goto flash

        ; Subroutines
        convert1
                bcf flags, 0        ; Clear reflecting flag.
                bsf portc, 4        ; Headlamps on.
                call shortdelay
                call adread
                movf adresh, w      ; Read 8-bit result into w.
                movwf reflect       ; Reflected light.

        ; Writing data EEPROM (optional).

                bcf portc, 4        ; Headlamps off.
                call shortdelay
                call adread
```

```
; Writing data EEPROM (optional).

        movf adresh, w       ; Read 8-bit result into w.
        subwf reflect, f     ; Reflected - background.
        btfss status, 0      ; Reflected >= background.
        goto convert1        ; Spurious readings - try again
        movlw 19h            ; Minimum significant difference.
        subwf reflect, w
        btfss status, 0
        return               ; No reflection detected.
        bsf flags, 0         ; Reflection - set flag.
        return               ; Reflection detected.

adread
        movlw 050h           ; Select clock 1/16 Fosc.
        movwf adcon1
        movlw 01h            ; Left just.,Voltage ref is supply.,
                             ; channel 0, enabled.
        movwf adcon0
        call delay           ; To sample voltage.
        bsf adcon0, 1        ; Start conversion.
waiting
        call delay
        btfsc adcon0, 1      ; If conversion complete.
        goto waiting
        return

spin
        bsf portb, 6         ; Buzzer on.
        bsf portc, 5
        call delay
        bsf portc, 6         ; Reverse and turn.
        bcf portc, 7
        movlw 05h
        call longdelay
        bcf portc, 6         ; Stop.
        bcf portb, 6         ; Buzzer off.
        bcf portc, 5
        return

shortdelay
        movlw 060h           ; Sets length of delay.
        movwf delay1
        call delay
        clrf delay1          ; Restore delay1 to zero.
        return
```

Add the *delay* and *longdelay* subroutines as on p. 185 and complete the program with the *end* directive on the last line. If you want to monitor the data while testing or developing the program, insert the routines listed on pp. 127-128 at the points indicated in the listing. They can be deleted later after the program has been fine-tuned.

This programming section has concentrated on the analogue processing available in the 16F690. But this is not all that the *Scooter* can do if it is given additional sensors or actuators. For instance, add a sound sensor and program it to start roaming when you click your fingers. Or mount an infrared sensor at its front, pointing downwards as in the *Quester*. It can then be programmed to avoid crossing a black line. It can be trapped simply by drawing a continuous line around it. If it is biased to turn left (p. 170) it could be trained to follow a line or run a maze.

A more ambitious project is to build two *Scooters* and let them communicate with each other. This could be by flashing LEDs or by radio. Joint behaviour of two robots is something to explore.

Programming in PICBASIC

This section describes BASIC programs that are similar to the assembler programs in their action. Use these and the flowcharts as a guide to converting other assembler programs into BASIC. Where the assembler and PICBASIC programs have a more or less identical action we will not describe this again. Refer to the previous section. Important differences are described in this section.

Hello World!

The program of the PICBASIC version is listed opposite. It is clear when we compare this listing with the assembler listing on pp. 184-185, that the BASIC listing is much shorter. This is generally true of all BASIC versions of assembler programs. BASIC has built-in routines for complicated tasks, while assembler has to be told what to do, one small step at a time. But the assembled program in machine code is probably always shorter than the version generated by BASIC compiler.

```
' Scoot01.txt

Symbol Portb = $6
Symbol Portc = $7
Symbol Trisb = $86
Symbol Trisc = $87
Symbol Ansel = $11E
Symbol Anselh = $11F

      Poke Trisb, 0        ' Port B all outputs.
      Poke Trisc, 0        ' Port C all outputs.
      Poke Ansel, 0        ' Digital input/output.
      Poke Anselh, 0       ' Digital input/output.

' Program begins here.
      Poke Portb, 0
      Poke Portc, 0
      Pause 5000           ' Delay 5 s.

      Poke Portc, $10      ' Headlamps on.
      Pause 5000           ' Delay 5 s.

      Poke Portc, $80      ' Go forward.
      Pause 2000           ' Delay 2 s,

      Poke Portc, $20      ' Side lamp on.
      Poke Portb, $40      ' Buzzer on.
      Pause 2000           ' Delay 2 s.
      Poke Portc, 0        ' Side lamp off.
      Poke Portb, 0        ' Buzzer off.

   End
```

A noticable feature of this listing is the high proportion of 'Poke' statements. PICBASIC normally uses the 'High' and 'Low' commands for making an output pin high or low. These commands refer only to the bits of Port B. For various reasons, the *Scooter*'s PIC is already set up to use Port C for most output operations. To change it to using Port B would mean altering the wiring, and rather than do this we have used 'Poke'. Also, Port B in the 16F690 controller has only four bits. In the assembler version we are able to set or clear individual bits but 'Poke' acts on all the bits of a port at the same time. There are ways of handling this, which are explained later.

Similarly, the 'Peek' command reads all bits at the same time.

Seeking the light

The PICBASIC version of this program has the same aim as the assembler version, but tackles it in a different way. In the assembler version, the *Scooter* spins round until the first time it detects a bright source of light. Then it moves towards it. In the BASIC version, the robot spins round for at least 360°, sampling as it goes. It finds which is the brightest of perhaps several sources. Then it moves towards the brightest source. The flowchart shows how it does this.

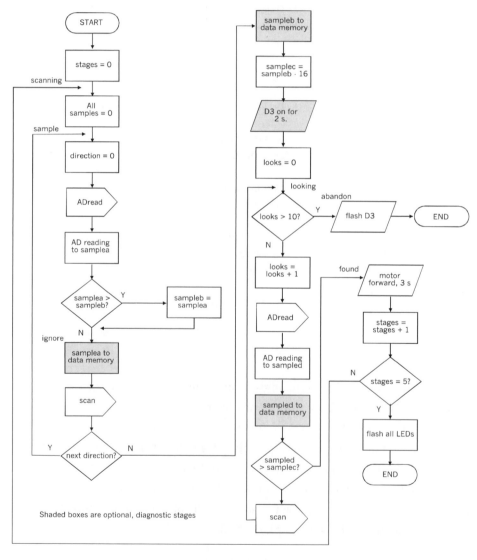

The BASIC version includes the routine for finding the maximum.

The *Scooter* is using one of its analogue-to-digital converters instead of its comparator. This means that it does not need to use pin 19 (AN0) for input from the voltage reference (VR1). We can just ignore this input or we can use it as an alternative input to a second AD converter. The PIC pin 19 (RA0/AN0) at F12 is connected to resistor R7 at E1. This makes LDR2 available as a light sensor on the left side of the robot.

The program includes a routine to find the brightest source. As it scans and samples, the successive samples read into *SampleA* are compared with *SampleB* and the greater is put into *SampleB*. At the end of the sampling phase *SampleB* is reduced slightly and stored in *SampleC*. This is so that sampling error is less likely to result in the robot missing the brightest source as it tries to find it again in the *looking* routine.

In the *looking* routine it scans again, comparing the input, *SampleD*, with *SampleC* until it finds a brighter (with a reasonable chance, the same?) source. Then it goes to the *found* routine and moves forward. The advantage of this program over the other version is that it will work at any overall light level. It does not depend on presetting a reference voltage. It samples input voltages and picks out the greatest. But there is a bug lurking here! Under very low light, with a not-very-bright source, the value of *SampleB* may be less than 16. Taking 16 away from this low value makes *SampleC* negative. It might be –3, for example, which is $fd in hex. However the compiler does not recognise this as a negative value. It is taken to be positive and $fd has the value 253 in decimal. This puts it way above any likely reading, and the true source will not be found. So, do not run this program in a coal cellar.

Here is the listing:

```
' Scoot03.txt

Symbol Portb = $6
Symbol Portc = $7
Symbol Adresh = $1E
Symbol Adcon0 = $1F
Symbol Trisb = $86
Symbol Trisc = $87
Symbol Adcon1 = $9F
Symbol Ansel = $11E
Symbol Anselh = $11F

Symbol SampleA = B2
Symbol SampleB = B3
```

```
    Symbol SampleC = B4
    Symbol SampleD = B5
    Symbol Looks = B6
    Symbol Finds = B7
    Symbol Direction = B8
    Symbol Stages = B9
        Poke Trisb, 0        ' Port B all outputs.
        Poke Trisc, 0        ' Port C all outputs.
        Poke Ansel, 3        ' Digital input except RA0 and RA1.
        Poke Anselh, 0
        Poke Portb, 0
        Poke Portc, 0
        Pause 4000
        Stages = 0
Scanning:
        SampleA = 0
        SampleB = 0
        SampleC = 0
        SampleD = 0
Sample:
        For Direction = 1 to 10
        Pause 500               ' Allow time to settle.
        Gosub ADread
        Peek Adresh, SampleA    ' Read result into SampleA.
        If SampleA < SampleB Then Ignore ' Find brightest.
        SampleB = SampleA
Ignore:
        Write Direction, SampleA
        Gosub Scan
        Next Direction
        Write 11, SampleB
        SampleC = SampleB - $10 ' Slightly smaller.
        Poke Portc, $20         ' D3.
        Pause 2000
        Poke Portc, 0           ' D3 off.
        Looks = 0
Looking:
        If Looks > 10 Then Abandon
        Looks = Looks + 1
        Pause 500
        Gosub ADread            ' Looking for bright light.
        Peek Adresh, SampleD
        Write 12, SampleD
        If SampleD > SampleC Then Found  ' Bright light found.
        Gosub Scan
        Goto Looking
```

```
Found:
      Poke Portc, $80           ' Forward.
      Pause 3000
      Poke Portc, 0        ' Stop.
      Stages = Stages + 1
      If Stages = 5 Then Finish
      Goto Scanning
Abandon:
      Poke Portc, 0        ' Stop.
Flash:
      Poke Portc, $20
      Pause 500
      Poke Portc, $0
      Pause 500
      Goto Flash
Finish:
      Poke Portc, $30           ' All LEDs.
      Pause 500
      Poke Portc, $0
      Pause 500
      Goto Finish
' Subroutines
ADread:
      Peek Adcon0, B0          ' Enable converter.
      Poke B0, 0
      Bit0 = 1
      Poke Adcon0, B0
      Poke Adcon1, $50    ' Clock 1/16 Fosc.
      Pause 5
      Peek Adcon0, B0           ' Start conversion
      Bit1 = 1
      Poke Adcon0, B0
Waiting:
      Pause 5
      Peek Adcon0, B0
      If Bit1 = 1 Then Waiting ' Test GO/DONE bit.
      Return               ' Done.
Scan:    Poke Portc, $40           ' Reverse.
      Pause 1000
      Poke Portc, 0        ' Stop.
      Pause 100      ' Allow time to halt.
      Poke Portc, $80           ' Forward.
      Pause 1000
      Poke Portc, 0        ' Stop'
      Return
End
```

When developing and testing this program, it is helpful to know the values of the input voltages. PICBASIC makes this simple. The short command 'Write 11, SampleB' puts the value of *SampleB* ito EEPROM at address 11. The PIC is returned to the programming board for this value to be displayed on the computer screen. In the sampling routine we can see the ten succesive values of *SampleA*, which gives us an idea of their variability. These write commands can be deleted when development is finished.

Avoiding obstacles

This has much the same structure as the assembler version:

```
' Scoot04.txt

Symbol Status = $3
Symbol Porta = $5
Symbol Portb = $6
Symbol Portc = $7
Symbol Intcon = $B
Symbol Adresh = $1E
Symbol Adcon0 = $1F
Symbol Trisa = $85
Symbol Trisb = $86
Symbol Trisc = $87
Symbol Adcon1 = $9F
Symbol Ansel = $11E
Symbol Anselh = $11F

Symbol Reflect = B2
Symbol Backg = B3
Symbol Flag = BIT8


        Poke Intcon, 0     ' Disable interrupts.
        Poke Porta, 0
        Poke Portb, 0
        Poke Portc, 0
        Poke Trisb, 0      ' Port B all outputs.
        Poke Trisc, 0      ' Port C all outputs.
        Poke Ansel, 3      ' Digital input except RA0 and RA1.
        Poke Anselh, 0
        Pause 5000
```

```
Scooting:
     Pause 100
     Flag = 0
     Poke Portc, $80          ' Motor forward.
     Pause 2000
     Poke Portc, 0            ' Stop motor.
     Gosub Convert1
     Pause 100
     Write 4, B1
     If Flag = 1 Then Spin
     Goto Scooting

' Subroutines

Convert1:
     Flag = 0                 ' Clear reflecting flag.
     Poke Portc, $10          ' Headlamps on.
     Pause 1000
     Gosub ADread
     Peek Adresh, Reflect     ' Read result into Reflect.
     Write 1, B2
     Poke Portc, 0            ' Headlamps off.
     Pause 1000
     Gosub ADready
     Peek Adresh, Backg       ' Read result into Backg(round).
     Write 2, B3
     Reflect = Reflect - Backg 'Subtract background light.
     If Reflect < 0 Then Convert1 ' Spurious result.
     If Reflect < 25 Then Done  ' No reflection detected.
     Write 3, B2
     Flag = 1         ' If reflection detected.
Done:    Return

Spin:
     High 6                   ' Buzzer on.
     Poke Portc, $50          ' Reverse and turn. + D3
     Pause 5000
     Poke Portc, 0            ' Stop.
     Low 6                    ' Buzzer off.
     Return

ADread:
     Peek Adcon0, B0          ' Enable converter.
     Poke B0, 0
     Bit0 = 1
```

```
        Poke Adcon0, B0
        Poke Adcon1, $50        ' Clock 1/16 Fosc.
        Pause 5
        Peek Adcon0, B0         ' Start conversion
        Bit1 = 1
        Poke Adcon0, B0
Waiting:
        Pause 5
        Peek Adcon0, B0
        If Bit1 = 1 Then Waiting ' Test GO/DONE bit.
        Return                  ' Done.

    End
```

6.2 The Android

The *Android* looks like the sort of robot featured in so many movies and TV plays. But although its appearance is traditional, its construction is novel.

The body is built from two low-density and easy-to-work materials. The body panels are cut from foam board. This can be done using a sharp craft knife and steel ruler. The joints are strengthened with balsa wood. The structure is held together with craft glue. The resulting body is remarkably rigid and strong, yet few tools and few skills are required to put it together.

Specification

Runs on 6 V battery.
Lightweight body.
Three wheels, 2 rear drive wheels, 1 motor-turned steering wheel at front.
PIC 16F690.
Sensors: light dependent resistors, push button input,
Actuators: Drive motor, steering motor, arm motor, bleeper, LEDs, speaker.

Programs

> Scissors, paper, stone
> Song and dance
> Light seeking
> Avoiding obstacles

Emphasising the modular approach of this book, many of the circuit units in this robot are the same as or slight modifications of corresponding units used by some of the other robots.

Mechanics

Start by building the chassis. It needs to be rigid and able to bear the strain of supporting the motors, steering gear, the body and its contents. Make it from plywood, composition board, or our favourite material, 3 mm expanded PVC sheet. The drawing overleaf shows the dimensions of the chassis, with cut-aways for the wheels, but the exact sizes and shapes depend on the motors and wheels you use.

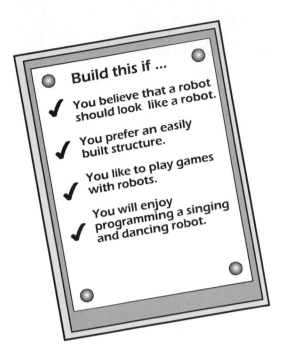

Build this if ...

✓ You believe that a robot should look like a robot.

✓ You prefer an easily built structure.

✓ You like to play games with robots.

✓ You will enjoy programming a singing and dancing robot.

The drive wheel assembly consists of a 6 V motor with gearbox and an output shaft on either side. The wheels are approximately 56 mm diameter. We used the same assembly as in the *Scooter* (pp. 169-170) but with the shafts cut shorter so that the wheels are 104 mm apart (centre to centre).

The steering wheel assembly consists of a pair of Lego® wheels clipped to a special brick, as in the *Scooter*. The steering gearbox is built from Meccano® parts. The motor with gearbox has a worm gear on its 4 mm diameter output shaft. This meshes with a 57-tooth gear. An alternative steering mechanism is discussed on pp. 224-226.

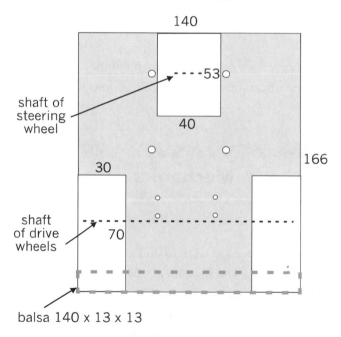

140

shaft of steering wheel

- - - - 53 ○

40

166

30

shaft of drive wheels

70

balsa 140 x 13 x 13

The chassis panel, showing the areas cut out for the wheels.

A view of the chassis from below shows how the wheels are within its rectangular area. This is so that the wheels will not be easily visible when the *Android* body is in place.

The splayed ends of the aluminium tubing (see lower photo) are visible in this view.

A rear view of the chassis with the wheels and motors installed. The chassis panel has a ledge projecting at the rear. A length of balsa is glued to this later, to support the body.

A front view of the chassis shows the steering gearbox and the steering wheel assembly. The gear wheel is on a standard Meccano 'triangular' section shaft. The lower end of this is a tight fit into a length of 5 mm diameter aluminium tubing. The central hole of the brick is drilled out to 5 mm diameter. The lower end of the tubing passes through this hole. The end of the tubing is cross cut for about 6 mm into four sections which are splayed outward between the four studs of the brick, to hold the brick on the tubing.

A lever is bolted to the gear wheel. The lever has a long bolt projecting downward to engage with the levers of the limit switches. These are mounted on a small platform of PVC sheet bolted to the frame of the steering mechanism. This photo was taken before the switches were wired up.

This view, from the right, shows the worm gear mechanism and the way the two limit switches are positioned. The steering wheels are able to swivel approximately 15° to either side of straight ahead.

The limit switches will be wired in parallel between the PIC input pin (channel RA5) and the 0 V line. Input goes low when either switch is closed.

Having built the assembly from Meccano and found that it works, you may decide to buy spares to replace the parts used. Or you may prefer to dismantle the Meccano version and build a new version, using aluminium stock and metal or plastic gears from other sources.

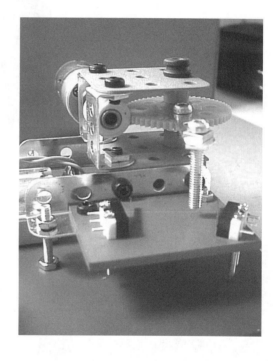

While assembling the chassis look ahead to the next stage to work out how the body is to fit on to and be supported by the chassis. The drive and steering mechanisms will be housed in the 'feet' of the body. There must be room for them to operate without touching against the inside of the body. Also it must be possible to remove the body to make alterations and adjustments to these mechanisms.

Before going on to build the body, start work on the electronics and programming of the drive and steering motors. It is easier to do this now and check their operation while the chassis and its mechanisms are easy to get at.

The body is built from Foam Board and 13 mm square balsa. There are no limits to the fancy designs but we decided to keep it simple. Hence the boxy *Android*. Use a very sharp blade for cutting, as the inner foam of the board tends to rip away from the surface films. Take care when cutting near to the corners of a panel as the same thing may happen here. Apart from taking these precautions, the board is simple to handle.

The height of the robot is a matter of preference. If it is too short, it will not look like the conventional android. If it is too tall, it may tend to topple over. Our prototype is 360 mm tall. When planning it, and the layout of the items inside it, keep the centre of gravity as low as possible, and above a point that is well inside the triangle formed by the contact points of the three wheels. The chassis with the drive and steering motors and gearing are at the lowest possible level, so this helps keep the centre of gravity low. Arrange to have that other heavy item, the battery, as low down as possible.

The Foam Board is 5 mm thick and some of the cut edges are visible. They are shaded in grey in the drawing.

The 80 × 30 cut-away in the sides is to allow extra width for the rear wheels. The panel shown in dashed lines is glued on to cover the cut-away.

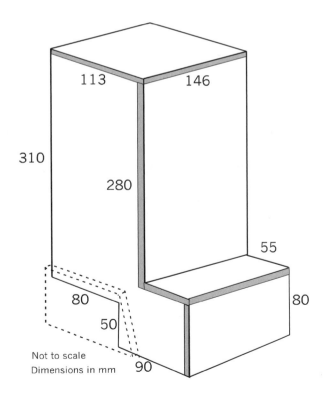

Not to scale
Dimensions in mm

The drawing illustrates one way of putting the body together. There are several other ways, depending on how you want the robot to look.

This is the body from the rear, showing the balsa strips glued in the angles. The two strips low down on the sides are to suppot the shelf that carries the battery and the main circuit boards. To allow easy access for checking, the shelf is not glued to these supports.

There are two cross strips at the top and bottom. The lid is slightly longer than the distance between the strips. To insert it, the lid is slid up behind the top strip, then drops down behind the bottom strip. Blocks of balsa stop it dropping too far.

From the front, the *Android* looks imposing with its dark blue board and the white edges of the board. The next step is to give it some arms.

Arms

Cut two of these from Foam Board. If you intend to program the *Android* to play *Scissors, Paper, Stone*, one of these arms is attached to the shaft of a stepper motor. This is an example of one way of providing moving parts for the *Android*. In the prototype the right arm swings up to indicate the 'call'. The other arm is just glued in place, but this too could be moved by its own stepper motor.

An arm is cut from a rectangle of Foam Board about130 mm by
40 mm. The movable arm has three holes in the 'shoulder' for bolting
on the pulley wheels that attach the arm to the shaft.

The stepper motor is bolted inside the body with its output shaft projecting through a hole at the shoulder position. The arm is clamped between a pair of 28 mm diameter pulleys from a Tamiya® pulley kit. They are secured by three bolts. A bush from the same kit fits tightly into the hub of the inner pulley and this bush fits firmly on to the drive shaft of the motor.

When the arm turns to one of its fixed
positions, the notch at its upper end reveals
one of several symbols. These could indicate
the robot's play in the *Scissors, Paper, Stone*
game. There are several possible applications
for this feature, particularly in games
programs.

Other peripherals

The speaker is a miniature piezo speaker, diameter about 30 mm. It is mounted inside the body just below the head. Drill a cluster of 2 mm holes in the body wall to allow sound to escape and glue the speaker by its rim to the wall, behind the cluster. If necessary, cover the perforated area outside by gluing a circle of fabric over it.

Drill two holes in the head to take the pair of LEDs that are its 'eyes'. With Foam Board the hole can be a little smaller than the diameter of the LED. For 5 mm LEDs, holes 4.5 mm in diameter gave a good push fit. If they are to spare, use the two LEDs 1 and 2 taken from the *Scooter*.

Shopping list — mechanical

5 mm Foam Board, 1 sheet.

3 mm expanded PVC board or plywood approx. 140 mm X 166 mm.

Balsa, 13 mm X 13 mm X 2 m.

Craft glue (Tacky glue).

DC motor and gearbox with output shaft on both sides.

DC motor and gearbox, or stepper motor for steering.

DC stepper motor for arm.

Parts for worm drive, if used.

Pair of wheels, with tyres, about 60 mm diam.

Metal tubing to fit into wheels and fit over drive shaft of motor.

Pulley wheels (2) about 25 mm diam for arm.

Insulating tape, coloured, for decorating robot body.

Nuts and bolts, mainly M3, bolts 6 mm and and 10 mm long.

Electronics

This robot is a demonstration of how circuit modules belonging to one robot can be used in another. Either build them again for the new robot, or take the old robot apart and use its circuit boards for the new one. We adopted the second course for the *Android*. Having already built the *Scooter*, we took its controller board, motor control board, and switching board to put into the *Android*. We also took the motor control board from the *Quester*, so all the main parts of the *Android* came from other robots and needed only to be connected together.

The *Android* has three motors: drive, steering, and arm. The arm motor is needed for moving the right arm when playing *Scissors, Paper, Stone*. The left arm does not move in our version, though this could have a motor too.

The third motor, M3, which is used for steering, is rated to run on 4.5 V to 18 V. On a 6 V supply it runs slowly but fast enough. The H-bridge output is 1.4 V below the input so this motor is getting only 4.6 V. This motor is controlled by a motor control board like that from the *Scooter*, which is for a single motor. The other two motors use the double control board from the *Quester*.

The entire system operates on 6 V, provided by a battery of four alkaline cells. The power switch S1 is mounted on the right body wall, near the feet, checking that it does not obstruct the rotating arm.

Controller board

This is the same as the controller board of the *Scooter* (p. 174) but has three pull-up resistors added to it. These are 10 kΩ resistors soldered between E2 and F2, between E3 and G3, and between E4 and H4.

The controller board is bolted to the shelf on the right and close to the front (see photo overleaf) where it is easy to remove and replace the PIC.

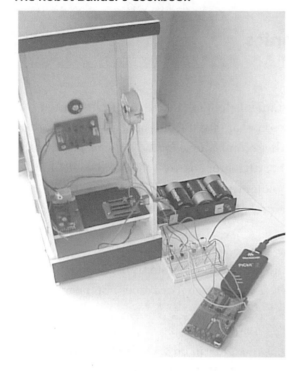

Before wiring up the board, the programming of the stepper motor of the arm was developed and tested. The motor is on the right-hand wall, with wires leading to the prototype motor control board (still breadboarded at this stage). The breadboard receives input from the PIC, which is on the PICkit 2 development board.

A closer view shows the shelf with the controller board at front right and the switching board on the left. At the back right corner, the shelf is cut away to allow wires for the drive and steering motors to pass down to the chassis beneath. The motor control board for the stepper will go on the shelf at back left. The battery (4 × AA) will go in the central space on the shelf.

The power switch is on the right wall just above the controller board. The motor control board for the drive and steering motors is mounted on the wall behind (from this view) and just above it is the speaker. The wires from this pass through a cable clip, which supports the 100 µF capacitor.

With all the essential circuit modules, except for one, taken ready-made from other robots there would seem to be little to do but join them together. But first there may be a problem to be attended to. The steering motor H-bridge does not provide a full 6 V supply so the motor may run too slowly or not at all. An interfacing circuit may be needed. The interface board and the steering motor are then supplied with a higher voltage (say, 9 V). which needs a separate power switch.

The alternative is to think again and shop around for a motor (perhaps a stepper motor) that *will* run on 4.6 V. In fact, a stepper could be better because it does not take its supply from an H-bridge so can obtain the full 6 V from the battery.

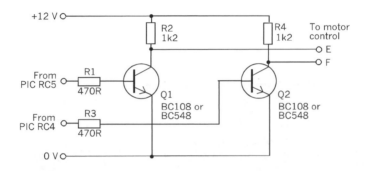

The interfacing circuit comprises two transistor switches. Almost any type will do. In the prototype they are a couple of BC108 transistors that have been unused in the spares box for a few years, but the more recent BC548 is equally suitable.

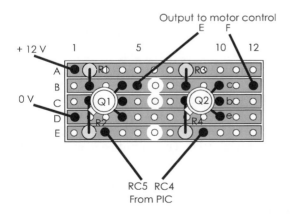

On the interface board, the connecting wires are soldered directly into the holes. Their other ends are stripped for inserting into the sockets on the controller and motor control boards. The power supply is taken from spare sockets on these boards.

The transistor switches of the interface are inverters, so the signal from the controller is inverted when it arrives at the motor. If the motor turns in the wrong direction, simply reverse the E and F connections between the interface and motor control boards.

Stepper motor control board

This interfaces the controller with the stepper motor inputs and consists of four transistor switches. The schematic of one switch is on p. 99. The circuit board layout is below. This board is run on the same supply as the controller (6 V), or its supply could be at higher voltage if the motor needs it.

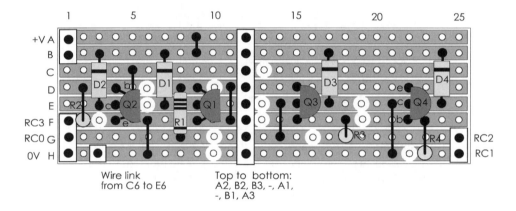

The stepper motor (M3) control board.

The board is most conveniently located at the back right of the shelf, next to the controller board.

Speaker drive

A piezo speaker can be driven directly from a PIC output. The circuit is given on p. 90.

If the sound is not loud enough, use a transistor switch to drive an 8 Ω speaker. The circuit has the speaker in its collector circuit and needs a protective diode (see the drawing on the right, p. 91).

With a 6 V supply and an impedance of 8 Ω, the current has a maximum of 750 mA. Build the switch on a small board with no resistor in series with the speaker and a base resistor of 470 Ω. Alternatively, and more conveniently, use Q3 on the switching board, but replace R5 with a wire link.

Off-board connections

The table below lists the connections for the power supply and for the motors. Make and test these first.

FROM		Function	TO	
Board	Socket		Board	Socket
Controller	C1	Positive supply	Switching	A1
Controller	D1	Positive supply	M3 control	A1
Controller	E1	Positive supply	Power switch S1*	
M3 control	B1	Positive supply	M1/M2 control	C2
Power switch S1 (common) *		Positive supply	Battery positive	
Controller	C12	0 V line	Switching	C1
Controller	D12	0 V line	M3 control	H1
Controller	E12	0 V line	Battery negative	
M3 control	H3	0 V line	M1/M2 control	E2
Controller	N12	M1/2 control A	M1/M2 A	G2
Controller	N1	M1/2 control B	M1/M2 B	H2
Controller	J1	M1/2 control C	M1/M2 C	I2
Controller	I1	M1/2 control D	M1/M2 D	J2
Motor control	G20	Motor output A	Motor M1 terminal*	
Motor control	H20	Motor output B	Motor M1 terminal*	
Motor control	I20	Motor output C	Motor M2 terminal*	
Motor control	J20	Motor output D	Motor M2 terminal*	

Next make the connections for the arm stepper motor as in the table below.

FROM		Function	TO	
Board	Socket		Board	Socket
Controller	I12	M3	M3 control A1	G1
Controller	J12	M3	M3 control A3	H25
Controller	K12	M3	M3 control B1	G25
Controller	K1	M3	M3 control B3	F1

The connections to the motor are as listed on the stripboard diagram on p. 220, using an 8-way plug and socket.

Finally, make the connections to the other off-board components.

FROM		Function	TO	
Board	Socket		Board	Socket
Controller	H1	Button	Button terminal*	
Controller	G12	LDR1	Switching	F1
Controller	F12	Comp. 1 ref.	Switching	B1
Controller	L12	D1/D2	Switching	B22
Controller	M12	Speaker	Speaker terminal*†	
Controller	H12	Bleeper	Switching	A22
Controller	F1	Steering limits	Limit switches	

*The other terminal of the button and the speaker are connected to 0 V. The anode terminals of D1 and D2 are connected to the positive supply. †The speaker may be switched by Q2 on the switching board.

Stepper steering

Alternatively, the off-board connections and the I/O table for this are on pp. 232-234.

PIC I/O

For later reference, the table below lists I/O connections to the PIC. A photocopy of this table mounted on card is handy to have on the workbench while assembling and testing the circuit modules.

Port	Name	IC pin	I/O	Connected to	0 =	1 =
A	RA0	19	AnI	Reference voltage, Comparator 1(C1IN+)		
	RA1	18	AnI	Forward light sensor, Comparator 1 (C12IN-)		
	RA2	17	I	Program select		
	RA3	4	I	Button switch	Pressed	Not
	RA4	3	I	Spare (input only)		
	RA5	2	I	Steering limit switches	Limit	Free
B	RB4	13	O	Headlamps	Off	On
	RB5	12	O	Speaker		
	RB6	11	O	M1, Drive motor A	Forward	Reverse
	RB7	10	O	M1, Drive motor B	Reverse	Forward
C	RC0	16	O	M3, Arm motor A1	High	Low
	RC1	15	O	M3, Arm motor A3	High	Low
	RC2	14	O	M3, Arm motor B1	High	Low
	RC3	7	O	M3, Arm motor B3	High	Low
	RC4	6	O	M2, Steering motor C	To left	To right
	RC5	5	O	M2, Steering motor D	To right	To left
	RC6	8	O	Bleeper	Off	On
	RC7	9	O	Spare		

The entries are grouped by I/O port and list all the channels available on the PIC16F690. In the fourth column, the I or O indicates whether the channel is to be configured as an input or an output. AnI is an analogue input.

The table shows the settings of motor M1 A and B lines that provide forward/reverse/off control. To stop the motor make both 0 or both 1. If the motor turns in the wrong direction reverse the connections to the motor control board or to the motor. The same applies for lines C and D of motor M2.

M3, which moves the arm, is a stepper motor with four control inputs.

Steering by stepper motor

The ordinary DC motor, M2, is replaced by a stepper motor of the same type as used for the arm. It has the advantage of more accurate and probably faster changes of direction. It might be more compact, too.

FROM		Function	TO	
Board0	**Socket**		**Board**	**Socket**
Controller	C1	Positive supply	Switching	A1
Controller	D1	Positive supply	M3 control	A1
Controller	E1	Positive supply	Power switch S1*	
M3 control	B1	Positive supply	M2 control	A1
M2 control	B1	Positive supply	M1 control	C1
Power switch S1 (common) *		Positive supply	Battery positive	
Controller	C12	0 V line	Switching	C1
Controller	D12	0 V line	M3 control	H1
Controller	E12	0 V line	Battery negative	
M3 control	H3	0 V line	M2 control	H1
M2 control	H3	0 V line	M1 control	E2
Controller	N12	M1 control	M1 A	G2
Controller	N1	M1 control	M1 B	H2
Controller	J1	M2 control	M2 A1	G1
Controller	I1	M2 control	M2 A3	H25
Controller	L1	M2 control	M2 B1	G25
Controller	M1	M2 control	M2 B3	F1
Controller	I12	M3 control	M3 A1	G1
Controller	J12	M3 control	M3 A3	H25
Controller	K12	M3 control	M3 B1	G25
Controller	K1	M3 control	M3 B3	F1

The arrangement of motor control board is amended to a single H-bridge board for M1, the drive motor. This would be like the board used in the *Scooter*, p.175. The new steering motor, which is referred to as M2, needs a second control board like that used for M3.

PIC I/O (Stepper steering)

For later reference, the table below lists I/O connections to the PIC. A photocopy of this table mounted on card is handy to have on the workbench while assembling and testing the circuit modules.

Port	Name	IC pin	I/O	Connected to	0 =	1 =
A	RA0	19	AnI	Reference voltage, Comparator 1(C1IN+)		
	RA1	18	AnI	Forward light sensor, Comparator 1 (C12IN-)		
	RA2	17	O	Bleeper	Off	On
	RA3	4	I	Button switch	Pressed	Not
	RA4	3	I	Steering limit, L	Limit	Free
	RA5	2	I	Spare (WPU)		
B	RB4	13	O	Headlamps	Off	On
	RB5	12	O	Speaker		
	RB6	11	O	M1 Drive motor, A	Forward	Reverse
	RB7	10	O	M1 Drive motor, B	Reverse	Forward
C	RC0	16	O	M3, Arm motor, A1	High	Low
	RC1	15	O	M3, Arm motor, A3	High	Low
	RC2	14	O	M3, Arm motor, B1	High	Low
	RC3	7	O	M3, Arm motor, B3	High	Low
	RC4	6	O	M2 Steering motor, A1	High	Low
	RC5	5	O	M2 Steering motor, A3	High	Low
	RC6	8	O	M2 Steering motor, B1	High	Low
	RC7	9	O	M2 Steering motor, B3	High	Low

The entries are grouped by I/O port and list all the channels available on the PIC16F690. In the fourth column, the I or O indicates whether the channel is to be configured as an input or an output. AnI is an analogue input.

There are two stepper motors: M2 is the steering motor and M3 moves the arm. Both have four control inputs.

Shopping list — electronic

Controller board:
> R1 - R3 resistors 10k (3 off)
> Other components as on p. 182.

Motor control board (double):
> Q1, Q3, Q5, Q7 BC639 npn transistor (4 off)
> Q2, Q4, Q6, Q8 BC640 pnp transistor (4 off)
> socket strip or PCB plugs 4-way (2 off)
> stripboard 16 strips x 21 holes

Motor control board (single, for DC motor steering):
> Q1, Q3 BC639 npn transistor (2 off)
> Q2, Q4 BC640 pnp transistor (2 off)
> socket strip, 2, 3, and 6 sockets
> stripboard 8 strips x 21 holes

Motor control board (single, for each stepper motor):
> Q1-Q4 BC639 npn transistor (4 off)
> R1-R4 resistor, 470R (4 off)
> D1-D4 1N4001 or similar rectifier diode (4 off)
> socket strip, 1 2 (2 off) 3, and 8 sockets
> stripboard 8 strips x 25 holes

Switching board:
> As on p. 182

Off-board components:
> D1 - D2 5 mm light emitting diodes, ultra bright (2 off)
> LDR1 light dependent resistor (ORP12 or similar)
> S1 SPST mini toggle switch
> battery holder, 4 x AAA or 4 x AA, with wire or stud terminals
> PP3 type battery connector (if battery box has stud terminals)
> AAA or AA NiMH rechargeable cells (4 off) (preferred)
> S2 SPST push-to-make push-button
> S3 Microswitch (2 off if steering by DC motor)

Miscellaneous:
> Strip of Velcro™ sticky Back®
> Single stranded connecting wire (to fit sockets)
> Solder

Programming

Having 'borrowed' some of its circuit boards from *Scooter*, the *Android* can run the same programs. These are *Hello World!* (pp. 183-186), the light-seeking program (pp. 190-194) and the object avoiding program (pp. 194-200). The I/O allocations of the *Android* differ from those of the *Scooter* so the listings must be amended:

- Bleeper has moved from RB6 (pin 11) to RC6 (pin 8).

- The headlamps, D1 and D2, have moved from RC4 (pin 6) to RB4 (pin 11).

- The side lamp, D3 at RC5 (pin 5) is not installed on the *Android*, but there is a spare channel at RC7 (pin 9) that could be used.

- Motor M1 has moved from RC6/7 (pins 8/9) to RB6/7 (pins 10/11).

The main difference is that *Android* has a motor M2 for steering, instead of the eccentric pulley. Where the listing shows the robot changing its direction of motion by backing a short distance, the instruction to reverse the motor M1 is omitted. Instead, motor M2 is programmed to alter the direction of steering.

Going straight

The first thing to be done in almost every program is to turn the steering wheel so that the robot will run straight ahead. The wheel may have been left at an angle at the end of the previous session, so the robot begins the new session, not knowing which way it is heading. This is not a problem for robots with tank steering.

The program turns the steering wheel in one direction until it triggers the limit switch on that side. It then knows where the wheel is pointing and turns it back by a fixed amount to centre it. There are two listings (both overleaf) for this routine, one for use with a DC motor and the other for a stepper motor.

The DC motor routine relies on timing the return to the central position, so will require fine-tuning and, even then, may not be accurate. The stepper motor routine is very accurate, provided that the limit switch is accurately placed.

On the subject of limit switches, steering by DC motor is less precise than steering by stepper. With a DC motor there is cumulative error and there must be two limit switches, left and right, to prevent it from turning too far in either direction. With a stepper motor we need only one limit switch for initially centreing the wheel. After that the steps are counted every time the wheel is turned. Its position is held as the value of a variable and the robot always knows which way it is heading.

Below is the subroutine for use with a DC motor. It calls on the *delay* and *longdelay* subroutines. It is called at the beginning of the main program:

```
straight
     btfss porta, 4      ; Left MS closed?
     goto centre         ; If closed.
     bcf portc, 5        ; Turn wheel left.
     bsf portc, 4
     call delay
     bcf portc, 4        ; Stop turning wheel.
     goto straight       ; Until fully left.
centre
     bcf portc, 4
     bsf portc, 5
     movlw 030h          ; This value needs tuning.
     call longdelay      ; Wheel right,to centre
     bcf portc, 5        ; Stop turning.
     return
```

The listing opposite is complete, except for the initial statement and equates. The equate list includes four variables, mask, *pointer, count* and *times*. Immediately before the start of the program there is a look-up table, *codes*. This table lists the codes that are sent to the motor to make it turn one step to the left or right. The pointer points to the current code the motor has reached. The next code read from a table will be the correct one for continuing to turn in the given direction.

There are only four codes so a pointer should return to code 0 after pointing to code 3 (or to 3 after 0 wen counting down). This is easily arranged by using a mask, value 0000 0011 in binary and ANDing this with the pointer to read only the lower two bits. The actual value in the pointers runs up to 255 before resetting but this is ignored by ANDing out the lower two digits. Note that the beeper is moved from RC6 to RA2 when steering is done by stepper motor.

```
goto start
      org 04h
      goto start
codes
      addwf pcl, f
      retlw 90h    ; high nybble of Port C.
      retlw 50h
      retlw 60h
      retlw 0a0h

start
      bcf intcon, 7      ; Disable inter-
rupts.bcf status, 5      ; Bank0.
      bcf status, 6
      bsf status, 5      ; Bank1.
      movlw 0fbh
      movwf trisa  ; Port A <2> is output.
      clrf trisb          ; All outputs.
      clrf trisc          ; All outputs.
      bcf option_reg, 7  ; Weak pull-ups.
      bsf wpua, 4         ; WPU on RA4.
      bcf status, 5       ; Bank2.
      bsf status, 6
      clrf ansel          ; For digital I/O.
      clrf anselh
      bcf status, 6       ; Bank0.
      bcf status, 5
      clrf porta
      clrf portb
      clrf portc

; Program begins here

      clrf pointer        ; Clear pointer.
      movlw 03h
      movwf mask
      movlw 010h
      call longdelay
      call beeps
      call straight ; Centre steering wheel.
      bcf portb, 6        ; Forward.
      bsf portb, 7
      movlw 0ah
      call longdelay
      bcf portb, 7        ; Stop.
      call beeps
```

```
      bsfportb,6  ;Reverse.
      bcf portb, 7
      movlw 0ah
      call longdelay
      bcf portb, 6          ; Stop.
flashem
      bsf portb, 4
      call delay
      bcf portb, 4
      call delay
      goto flashem

; Subroutines
delay
      decfsz delay0, f
      goto delay
      decfsz delay1, f
      goto delay
      return
longdelay
      movwf delayn
repeat
      call delay
      decfsz delayn, f
      goto repeat
      return
beeps
      movlw 05h
      movwf times
domore
      bsf porta, 2
      movlw 05h
      call longdelay
      bcf porta, 2
      movlw 05h
      call longdelay
      decfsz times, f
      goto domore
      return
straight
      btfss porta, 4  ; Left MS closed?
      goto centre            ; If closed.
      movlw 01h
      movwf count
      call cyclel ; Turn wheel left.
      call delay
```

```
        goto straight      ; Until fully left.
centre
        movlw 06h   ; For 6 steps (45 deg.).
        movwf count
        call cycler
        return
cycler
        movf pointer, w
        decf pointer, f     ; Next right.
        andwf mask, w
        call codes

        movwf portc        ; Step motor.
        call delay
        call delay
        decfsz count, f    ; Count steps.
```

```
        goto cycler
        return
cyclel
        movf pointer, w
        incf pointer, f     ; Next left.
        andwf mask, w
        call codes
        movwf portc
        call delay
        call delay
        decfsz count, f
        goto cyclel
        return

        end
```

Song and Dance Act

First the song. A routine for producing a square-wave tone is described on pp. 150-155. This produces a single burst of sound at a given frequency and of a given duration. But before it can do this it needs three values to work on. It also needs editing to produce the sound signal at the specified pin.

In the *Android*, the speaker is driven from Port B, channel RB5. Edit the listing to read `portb` instead of `portc`, `trisb` instead of `trisc`, and `portb, 5` instead of `portc, 0`. To obtain the pitch of Middle C, type in this list of values:

```
        movlw 024h
        movwf data1
        movlw 020h
        movwf datafinal
        movlw 02ah
        movwf lendata
        call playit
```

This is part of the listing of the main program so type `call playit` immediately after it. If you are trying this out as a demo, prevent the PIC from running on into the subroutines by typing a waiting loop such as:

```
        done   goto done
```

If you intend to sound the same tone several times in a program, the whole thing can be put into a subroutine.

If you want a sequence of two to four notes, repeat the list that number of times and call playit after each. For example, this routine plays C, C', G':

```
movlw 024h          ; Play C
movwf data1
movlw 020h
movwf datafinal
movlw 0ah
movwf lendata
call playit
movlw 010h          ; Play C'
movwf data1
movlw 02Fh
movwf datafinal
movlw 015h
movwf lendata
call playit
movlw 0ah            ; Play G'
movwf data1
movlw 029h
movwf datafinal
movlw 07eh
movwf lendata
call playit
```

With these settings the tones last for one or more seconds. The simplest way to make them shorter is to reduce the value in the outer loop counter from 5 to 3, or 2.

For a proper tune it is more practicable to use a look-up table. Each entry in the table holds three values: `data1`, `datafinal`, `lendata`. The table for the 3-note 'tune' above is given overleaf. It lists the three values for each note, one after another.

The subroutine for reading the table reads three values, using *pointer*, puts them in the three variables, and then calls *playit*. For the next note, *pointer* is again incremented three times, reading and transferring them each time, before calling *playit*. And so on until all notes have been played.

The listing for the 3-note sequence is:

```
yourtune
      addwf  pcl,  f
      retlw  024h           ; This is data1, first note.
      retlw  020h           ; This is datafinal, first note.
      retlw  02ah           ; This is lendata, first note.
      retlw  010h           ; This is data1, second note.
      retlw  02fh           ; This is datafinal, second note.
      retlw  054h           ; This is lendata, second note.
      retlw  0ah            ; This is data1, third note.
      retlw  029h           ; This is datafinal, third note.
      retlw  07eh           ; This is lendata, third note.
```

A subroutine such as this can be extended to many more notes, so building up a complete melody. First work out the values, using the method described on p. 150. Then work out the table.

Now for the dance! Controlling the motors is done by sending a sequence of codes to the drive motor M1 and the steering motor, M2. A stepper is better for steering as it can be programmed to turn quickly (but not too quickly, or it may lose some steps). The codes can be included in the look-up table at every fourth (forward/reverse/off) and fifth (left/right) place.

The dance codes start or stop the drive or steering motors, which continue running while the next note is played. When the note is finished the PIC goes to the look-up table to find out what to do next. The motors may be reversed, stopped or left unchanged. It can be left to continue its motion unchanged for several notes. Choreographing a robot is a fascinating exercise.

Scissors, Paper, Stone

This international strategic game is one that the *Android* loves to play. With clever programming it might even learn how to win.

This is a game for two players. They stand facing each other with their right hands behind their backs. They form their right hands to symbolise scissors, paper, or stone.

For scissors, the hand is flat with the second and third fingers spread apart. For paper, the hand is flat with fingers together. For stone, the hand is clenched in a fist.

On a given signal, the players swing their hands forward and upward to reveal their choice of scissors, paper or stone. If they have both chosen the same, the result is a draw but, if they have chosen differently, the winner is decided like this:

- Scissors wins against paper because scissors cut paper.

- Paper wins against stone because paper wraps stone.

- Stone wins against scissors because stone blunts scissors.

Such a simple game may seem trivial, but some people take it seriously. As recently as May 2005, it was reported on *BBC News* that the famous auction company Christie's won a £10.5m contract as the result of a single round of this game in a contest against Sotheby's.

The program is listed in full on pp. 240-245 and there are flowcharts on pp. 237-239. To run this game on the *Android*, the right arm must be driven by a stepper motor. Two sensors are required: the Program Select switch and the push-button. The state of the game is indicated by the eye LEDs (D1 and D2) and the buzzer. In this program, the Program Select is not for making a choice between different programs but for making a choice between two different playing strategies for the robot.

The procedure for the game as played with the *Android* is as follows:

1) Turn the Program Select switch on or off. If it is off, the *Android* plays randomly and it is a matter of chance if it wins or not. If it is on, the robot is programmed to learn its best play by experience.

2) Run the program. The eyes light up at half brightness.

3) When you are ready to begin, press and release the button. The eyes go dark, then flash once when the button is released.

4) The arm is probably already raised. Press and hold the button. The arm turns clockwise. This stage is timed. If you do not press the button soon after stage (3), the program goes to Stage (5), but with the arm not in the correct position.

5) When the arm is pointing vertically down, release the button.

6) The robot makes its choice and the beeper sounds, As the sound ends, the robot raises its arm to indicate its choice. You do the same.

7) The robot's choice can be read in the notch at its shoulder. Or judge by eye how much it has turned: 45° means scissors, 90° means paper, 135° means stone.

8) The robot can not detect what response you made, so it needs to be told if it has won. If it has won, press the button once. If not, or if the contest resulted in a draw, do nothing. After about 1 minute, the robot will lower its arm and it is ready for the next game.

9) To play again with the same robot strategy, press the button to return to step (6). The robot's strategy can be changed by re-setting the Mode Select switch first.

This is a long listing and it is noticeable that the major part of it is concerned with sequencing the input and output routines. The actual game, the *strategy* subroutine, is relatively short.

There is plenty of scope for adding to the program. For example, the robot could use the speaker and generate different sound signals at each stage of the game. Or it could be programmed to do a short Song and Dance act whenever it wins, and perhaps emit a dismal sound when it loses.

To take it further, there are many learning strategies that the robot can adopt. This program is structured to allow a new subroutine to be slotted in, replacing one of the existing subroutines. Try programming a different strategy, one that you think will be a winner.

The listing has the usual set of equates for Special Function registers. The general purpose registers are:

delay0, delay1, and *delayn* for the *delay* and *longdelay* subroutines.

ranval, bitn, and b*itm* for the random number generator.

count, holds 01, 10, or 11, later multiplied by 8 to give the number of steps the motor is to turn.

pointer, points to the codes in the *code* look-up table.

mask to mask out bits <7:2> of *randval,* leaving bits <1:0>.

waitime, for setting a time-out in the *ready* subroutine.

countcopy holds a copy of the inital value of *count* for subsequent use.

newcount holds 01, 10, or 11, which determine the robot's next response. If it wins, *newcount* is not changed. If it loses, *newcount* is changed from 01 to 10, from 10 to 11, or from 11 to a randomly selected value.

'FLUFFY' WAS PROGRAMMED TO DO JUST ONE THING...

A flowchart of the main program is shown opposite. It begins by running the random number subroutine repeatedly until you decide to start playing, and press the button. The effect of this is that, because the timing is never exactlty the same twice, the game is almost certain to begin with a different random number every time.

The listing continues with a double check on the button to prevent the controller running on at high speed as soon as the button is pressed. Then comes a sequence of four calls to subroutines. The eye LEDs are flashed once and then come the *ready* and *moveit* subroutines, charted on p. 238. These turn the arm to hang vertically downward, if it is not down already. For as long as the button is held pressed by the player, it calls the *moveit* subroutine, in which the *cyclel* subroutine advances the stepper motor by one step. If the button is released, the *ready* subroutine by-passes *moveit*. Either way, there is a countdown of *waittime*, after which the position of the arm can not be adjusted.

Returning to the main program, the eyes are flashed three times to announce that the *Android* is ready to play. To decide on its response the program calls the *strategy* subroutine, charted on p. 239. After flashing the eyes again, the subroutine reads the Program Select switch. It sends the PIC to *strategy1* if the switch is off. There it selects a random number and uses *mask* to pick out the two lowest bits. If these are '00', the process is repeated until the digits are '01', '10', or '11' (which are 1, 2, or 3 in decimal). These correspond to scissors, paper, stone, so the result is a random choice of the robot's play.

To translate this into moving the arm, the value (now in *count*) is multiplied by eight to give 8, 16, or 24. This is done by rotating *count* left three times. Eight steps of the motor turn it by 45°, so the value in *count* turns the motor by 45°, 90°, or 135°. The value is stored in *countcopy* for use later when the arm is being returned to its vertical position. The subroutine return to the main program with the bits (multiplied by 8) stored in *count*, to determne how far the arm is to be raised.

In the alternative mode of play, *strategy2,* the robot repeats a successful play for as long as it continues to be successful. The aim of the algorithm is to give it a slightly better than average chance of winning. The value in *newcount* is tested on entering the *strategy2* routine. On the first time round it will be zero and the PIC is directed back to *strategy1* to pick a random value and place it in *count*. If this is not the first game of a session, the value in *newcount* is other than zero. This is assigned to count and multiplied by 8 before returning to the main program.

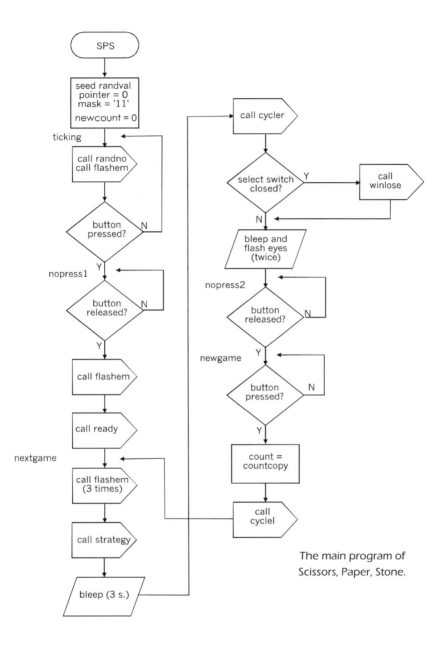

The main program of
Scissors, Paper, Stone.

The value in *newcount* may be one that has resulted in a win for the robot. It is repeating its successful play. However, if the robot lost the previous game, *newccount* is usually (but not always) given a different value in the *tryagain* routine (the last routine in the listing, p. 245). It will try a different response in the current game.

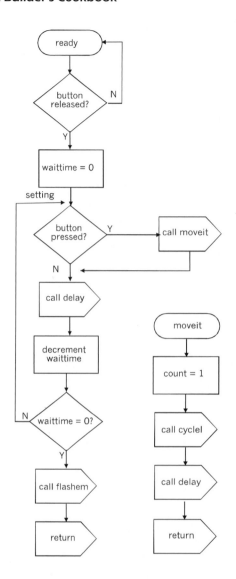

The *ready* and *moveit* subroutines, which prepare the robot for playing.

Following a return from the *strategy* subroutine, the PIC calls *cycler*. This subroutine (see top right of the flowchart on p. 237, and the listing on p. 243) turns the arm by the number of steps held in *count*. At each step it refers to the *codes* look-up table and sends the code to the stepper motor.

In *cycler*, the pointer is decremented at each step so the codes are read from the bottom of the table upward. In *cyclel*, used later to automatically lower the arm, the pointer is incremented at each step, and so reads the table from top to bottom.

The game is over now and the robot has either won or lost. When playing according to *strategy1* (switch not closed) the robot emits a bleep and flashes its eyes. It then waits for the player to press the button to initiate a new round.

Under *strategy2* (switch closed) it is sent to the *winlose* subroutine where it waits (at *scoring*) for the player to press the button if the robot has won. The length of wait is set by the value placed in *waittime*. The bleeper sounds while it is waiting. After checking that the button has been released, it copies *countcopy* into *newcount*. But this is eight times the original value so the bits 3 and 4 of *countcopy* are copied to bits 0 and 1 of *newcount*. This restores the correct value to *newcount*, ready for the next round.

Flowchart of the *strategy* subroutine.

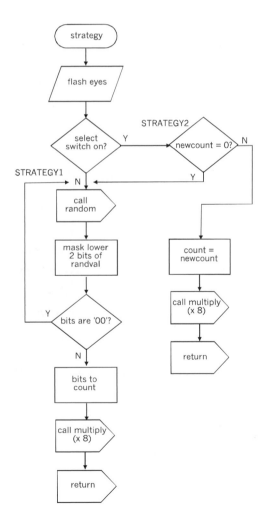

The PIC then waits (at *waiting*) for the button to be pressed while *waittime* is counted down to zero. If it is pressed the PIC returns to the main program with the winning count stored in *newcount*.

If it is not pressed,the program jumps to *result*, where *newcount* is incremented to give the robot a new response. However, if its previous response was 11, incrementing this gives a value that is outof range. In this case, *tryagain* selects a value at random. This may be the same as the previous value, but the chances are that it will be different.

In either mode, the program waits for the button to be pressed before it lowers the arm, using *cyclel* and counting down *count*. The program goes back to *nextgame*. After this, three eye flashes and a bleep signal the beginning of the next game.

After the initialising `list` statement and __CONFIG statement there is a list of equates:

delay0	equ 20h
delay1	equ 21h
randval	equ 22h
bitn	equ 23h
bitm	equ 24h
count	equ 25h
pointer	equ 26h
mask	equ 27h
waittime	equ 28h
countcopy	equ 29h
delayn	equ 2ah
newcount	equ 2bh

The program is as listed below and pp. 241-245:

```
goto start
        org 0004h
        goto start

codes
```

```
        addwf pcl, f
        retlw 09h               ; Codes in low nybble.
        retlw 05h
        retlw 06h
        retlw 0ah

start
        bcf intcon, 7           ; Disable interrupts.
        bcf status, 5           ; Bank0.
        bcf status, 6
        clrf porta
        clrf portb
        clrf portc
        bsf status, 5           ; Bank1.
        clrf trisb              ; All outputs.
        clrf trisc              ; All outputs.
        bcf status, 5           ; Bank2.
        bsf status, 6
        clrf ansel              ; For digital I/O.
        clrf anselh
        bcf status, 6           ; Bank0.
        bcf status, 5

; Program begins here

        bsf randval, 0          ; Seeding random numbers.
        clrf pointer            ; Clear pointer.
        movlw 03h
        movwf mask
        clrf newcount
ticking
        call flashem
        call randno
        btfsc porta, 3          ; Button pressed?
        goto ticking
nopress1
        btfss porta, 3          ; Wait for button release.
        goto nopress1
        call flashem            ; Flash eyes.
        call ready              ; To turn arm downward.
nextgame
        call flashem
        call flashem
        call flashem
        call strategy           ; Return with count set.
        bsf portc, 6            ; Bleeper on for 3s.
        movlw 010h
        call longdelay
        bcf portc, 6            ; Bleeper off.
        call cycler             ; To show choice.
        btfss porta, 2          ; If strategy2 used.
        call winlose
        bsf portc, 6            ; Ready to play?
```

```
          call flashem
          call flashem
          bcf portc, 6
newgame
          btfss porta, 3
          goto newgame
nopress2
          btfsc porta, 3
          goto nopress2
          movf countcopy, w ; Arm down.
          movwf count
          call cycle1
          goto nextgame

; Subroutines

delay
          decfsz delay0, f
          goto delay
          decfsz delay1, f
          goto delay
          return

longdelay
          movwf delayn
repeat
          call delay
          decfsz delayn, f
          goto repeat
          return

randno
          clrf bitn
          clrf bitm
          btfsc randval, 5   ;Getting n.
          bsf bitn, 0        ;If n = 1.
          btfsc randval, 6   ;Getting m.
          bsf bitm, 0        ;If m = 1.
          movf bitn, w       ;n to w.
          xorwf bitm, w      ;XOR m and n, result in w.
          addlw 0ffh         ;Set carry if w = 1.
          rlf randval,f      ;New random number in randval.
          return

flashem
          bsf portb, 4
          call delay
          bcf portb, 4
          call delay
          return

ready
          btfss porta, 3          ; Wait for button release.
```

```
        goto ready
        movlw 040h
        movwf waittime
setting
        btfss porta, 3          ; Press until arm down
        call moveit
        call delay
        decfsz waittime, f
        goto setting
        call flashem
        return                  ; Time expired.
moveit
        movlw 01h
        movwf count
        call cyclel             ; For one step.
        call delay
        return
cycler
        movf pointer, w
        decf pointer, f         ; Next position.
        andwf mask, w
        call codes
        movwf portc             ; Step the motor.
        call delay
        decfsz count, f         ; Counting the steps.
        goto cycler
        return


cyclel
        movf pointer, w
        incf pointer, f
        andwf mask, w
        call codes
        movwf portc
        call delay
        decfsz count, f
        goto cyclel
        return


strategy
        bsf portb, 4        ; Eyes on.
        movlw 05h
        call longdelay
        bcf portb, 4        ; Eyes off.
        btfsc porta, 2
        goto strategy1
        goto strategy2


strategy1                   ; Purely random.
        call randno
        movf mask, w
        andwf randval, w    ; Lowest two bits.
        btfsc status, z
        goto strategy1      ; If 00.
```

```
        movwf count
        call multiply      ; x 8.
        return

strategy2
        movf newcount, w
        btfsc status, z    ; First time through?
        goto strategy1     ; Yes, use random value.
        movwf count        ; No. Count = newcount.
        call multiply      ; x 8.
        return

multiply
        bcf status, 0      ; Clear carry bit.
        rlf count, f       ; count x 8.
        rlf count, f
        rlf count, f
        movf count, w
        movwf countcopy    ; Remember count x 8.
        return

winlose
        movlw 030h
        movwf waittime
        bsf portc, 6       ; Bleeper on - did robot win?
scoring
        btfss porta, 3     ; Wait for button release.
        goto scoring
        bcf newcount, 0    ; Copy  countcopy (x 8)
        btfsc countcopy,   ;  into newcount<1:0>.
        bsf newcount, 0
        bcf newcount, 1
        btfsc countcopy, 4
        bsf newcount, 1
nopress3
        call delay
        decfsz waittime, f ; Time expiring.
        goto waiting
        goto timeout
waiting
        btfsc porta, 3     ; Button pressed = robot won.
        goto nopress3
        return             ; With winning count in newcount.
timeout
        bcf portc, 6       ; Time up! Bleeper off.
        goto result        ; Not pressed - robot lost.
        return
result
        call rejectit      ; With losing count in newcount.
        bcf portc, 6       ; Loss registered, bleeper off.
        return
rejectit
        incf newcount, f   ; Next count to use.
```

```
        btfss newcount, 2 ; If it equals 4.
        return
tryagain
        call randno       ; To select a new random value.
        movf mask, w
        andwf randval, w
        btfsc status, z   ; If 00.
        goto tryagain
        movwf newcount    ; Newcount holds next response.
        return


        end
```

This completes the listing of the *Scissors, Paper, Stone* program. It is a 'bare bones' program, and takes up very little of the PIC's program memory. So there is plenty of scope and memory space for adding some 'frills', such as those suggested on p. 234.

6.3 A robotic toy

Some of us are not very good with our hands, or perhaps are bored with too much handicrafting and want to get on with the coding. One way round this is to start with a ready-made chassis. This may be purchased from a vendor of robot parts. Another course is to adapt a toy.

A warning here! If the toy is a new or fairly new one that is still covered by the manufacturer's Warranty or Guarantee, this may be made void the moment you start to prise open the toy or interfere with it in any other way. Better use a superannuated toy (with its owner's permission, of course!).

Specification

Based on an inexpensive or discarded toy.
Operating voltage depends on devices installed.
Any number of wheels, including none.
PIC 16F690.
Sensors and actuators as appropriate to the type of toy.

Programs described in this part:

> Variable motor speed
> Switching sound effects
> Gunshot sound effect

This project describes how a plastic toy car was given robotic capabilities. The same idea can be applied to a wide range of toys. Toys on wheels give the greatest scope for an interesting conversion. Instead of a toy vehicle, why not a toy duck?

Toys such as dolls and fluffy animals are more of a problem. Probably the best bet is to base a robot on a string puppet. Build a robot mechanism to pull the strings, equip it with a few sensors, then program it to walk, dance, or perform acrobatic stunts. Try to give it intelligence and the ability to learn.

So, if a toy car does not appeal, look for some other toy that would be fun to endow with robotic skills. For something different, how about a robot motor boat, or even a sailing boat? Teach it to sail close to the wind, to tack, and go about — all it needs is a tilt sensor and a wind direction sensor. If you are, or have been, a model railway enthusiast, there is a lot to be done in putting the system under intelligent robotic control.

────────────────────────────────────
Mechanics
────────────────────────────────────

When selecting a toy for this project check that it has space inside it to hold a battery and a few circuit boards. If it already has a battery holder, so much the better. The chunky style of toy vehicle designed for young children usually has plenty of room inside.

Also make sure that the plastic is not the kind that is liable to shatter when drilled. The construction of the toy should allow it to be opened up and put together again without damaging it. The car used in this project is held together with self-tapping screws, which are ideal for taking apart and reassembling the bodywork.

Our toy is a cheap car intended to be pushed along by hand. It has no drive motor and there is no steering. Its main asset is that it has press-buttons on one side. Pressing a button starts a sound effect playing and LEDs flashing. This feature is easily put under automatic control, as explained later.

The main mechanical tasks are to provide drive and steering motors and possibly stepper motors for other functions, such as raising the back of a tip-up truck. A large truck-type toy could provide a base for a ready-made robot arm, giving it mobility.

Build this if ...

✔ You believe that a robot should NOT look like a robot.

✔ You want to get on with the programming.

✔ You want something different.

Drive wheels

The car has four wheels, which turn in bearings on the sides of the body. We decided to use the rear wheels for driving the vehicle, and the front ones for steering, just as in a conventional automobile. The existing battery supply is from two type AA alkaline cells, giving 3 V. The motor/gearbox unit chosen as the drive motor runs on 3 V and has an output shaft on either side.

After considering various possibilities it was decided to push the wheels on to the ends of the output shafts. A short length of plastic sleeving pushed on to the shafts first was to make the wheels a tight fit. At this stage we discovered that it was impossible to remove the wheels from their original axle. So we gave up this idea and substituted Tamiya wheels, attached as described on p. 170. The hub caps of the wheels match those of the existing front wheels sufficiently well.

The drive motor is bolted to the floor of the bodywork and its shafts are just the right distance above the floor to pass freely through the existing bearings.

Steering

Steering is more of a problem because the front wheels are not able to turn from side to side. Also the front wheels are closely surrounded by mudguards so there is little space around them in which they could be turned if remounted. The answer to this is to raise the existing front wheels off the ground by a steering assembly similar to, but simpler than, that used in the *Android*.

The similarity is that the wheels are a pair of Lego™ wheels with tyres (left below). The brick has a 4.5 mm hole bored in it to take the long bolt. This is a 50 mm roofing bolt, 3/16" in diameter. The nuts that come with this bolt are square, so a portion of each stud is cut away (see close-up opposite) to allow the nut to fit between them. This prevents the nut from working loose.

The steering castor. There are two plastic collars on the bolt to set the distance between the wheels and the underside of the body.

The bolt passes upward through a 4.5 mm hole drilled in the lower body, centrally and just to the rear of the front wheels. The underside rests on the upper end of the collar and the length of this is such that there is a gap of about 5 mm between the front wheels and the ground. This is small enough not to be noticeable.

The four studs around the nut are cut away so that the nut fits neatly between them and can not turn.

The upper end of the castor bolt projects up into the front part of the body, under the bonnet. Washers are threaded on it, secured by nuts, so that the assembly turns freely yet does not wobble unduly. At the upper end of the bolt, a plastic pulley wheel is held between two nuts, tightened to prevent it from turning on the bolt (see below).

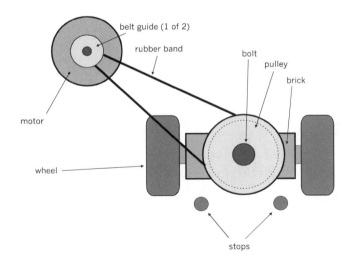

Looking down on the steering mechanism, we can see the upper end of the castor bolt projecting upward. At its upper end a 25 mm pulley is gripped between two nuts and a third nut is tightened against the upper one of these as a lock-nut.

The steering motor is a 3 V DC motor without a gearbox. It is small, only 20 mm in diameter and 25 mm long, so fits easily in the limited space inside the car. It does not have any means of bolting it in position, so it is held in a spring clip that is bolted to the side of the body. The clip is the kind used for holding a 9V PP3 battery. It grips the motor firmly yet the motor can be slid up or down to bring the 'pulley' on the motor's output shaft level with the pulley on the castor shaft.

There is no actual pulley on the motor shaft. The rubber band runs around the shaft itself. It is prevented from slipping off by two belt guides supplied as part of the Tamiya Pulley Set (see their instruction leaflet).

Two bolts project downward from the underside of the body. These are positioned as stops to prevent the castor from turning more than 45° on either side of 'straight ahead'.

A rubber band runs around the drive shaft and the pulley. When a brief pulse is applied to the motor, it rotates at high speed, turning the pulley/castor in one direction or the other. Almost instantly, it is prevented from turning further by one of the stops. The motor continues to rotate, but the band slips and the pulley remains directed at 45° to the left or right.

The steering action is much faster than that of the other steering mechanisms. The car zigzags along in a lively way. This vehicle is a good one to program for object avoidance and path following. Maybe we could teach it to run a slalom along a row of drinks cans.

More mods

For a toy vehicle, an amusing addition to the sensors is a downwardly-directed light sensor underneath, at the front. It could have its own light source, preferably infrared, or rely on ambient light. When the robot is running on a table there will be a change of light level as the front of the vehicle reaches the edge of the table. Program it to respond immediately by backing and turning before going forward again.

A sensor that detects the Earth's magnetic field is unusual but inexpensive. Installed in a toy vehicle (or in any of the other mobile robots) it provides a sense of direction. Precision is not high but good enough for simple navigation routines.

Electronics

The sound effects were an entertaining feature of the original toy, and we wanted to retain them for the robot. An effect is turned on by pressing a button on the side of the car. In the robotic version, they are switched on automatically by using relays.

Each button has a flexible plastic key-top, underneath which is a small (4 mm diam.) disc of conductive foam. The keytop adheres to the circuit board, which has two finger-like copper pads beneath each key-top, with a narrow gap between the pads. When the keytop is pressed the conductive disc bridges the gap between the pads and the sound effect is switched on.

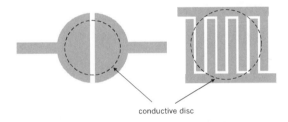

conductive disc

Two forms of copper pad used in push-to-make buttons of keypads.

The switching is put under the control of the PIC by wiring a relay in parallel with the push-button. A relay driving circuit is illustrated on p. 91. To connect the relay to the sound effects circuit board of the toy, carefully prise the key-top from the circuit board. With the power switched on, try short-circuiting the two pads together for an instant. This should initiate the sound effect and, if so, solder a wire to each pad and run the wires to the contact terminals of the relay. When the output from the PIC energises the relay its contacts close, connecting the pads and so turning on the sound.

Sounds add much realism to the robot's actions. If you have another spare toy with a different set of effects, it may be possible to install the circuit for these instead or as well. But first check that their operating voltages are compatible.

The electronic system includes a board for the controller and circuitry on or off board for the other devices that are installed. Both the drive and steering motors need to be reversible, so a transistor H-bridge or a relay board is needed for each motor. Browse the cookbook looking for other features to add to the robotic toy.

The sound effects run on 3 V from the built-in battery holder. A second battery holder holds two more AA cells (or AAA cells if space is short). It is wired in series with the built-in holder. The two batteries total 6 V for powering the controller board, the transistor switch that switches the relays, and the H-bridges that control the drive motor and steering motor.

Programming

Steering control

A pulse with the duration of a single call to *delay* is sufficient to turn the wheel fully in either direction. To steer straight ahead, first send a pulse to turn the wheel fully to right or left. Then send a shorter pulse to nudge the wheel in the opposite direction and into the forward (mid-way) direction. The pulse is obtained by calling *shortdelay* with 040h as the initial value of delay1. Adjust this value to suit your motor. The two pulses take only a fraction of a second, so the effect is a smooth change of direction.

Varible speed control

This routine has many applications. The idea is that power is supplied to a motor as a stream of pulses. The pulses are at the full supply voltage (or the voltage available from an H-bridge) but the ratio between the length of a pulse and the interval between pulses can be varied by programming. The shorter the pulse and, therefore, the longer the interval, the lesser the power delivered to the motor and the slower it turns. This way of controlling speed is preferable to varying the voltage. It makes the motor less likely to stall or fail to start at low speed.

The frequency of the pulses is around 100 Hz, so that the motor turns smoothly. The circuit may also be used for dimming a lamp or an LED.

The listing (pp. 254-255) is a demonstration of the routine and is useful for testing the action of the routine with a given motor or lamp.

The main program is short. It sets the values of three essential parameters of the rectangular waveform that drives the motor. In this demo they are typed into the program before compiling and running it. In an application they would be calculated by programming logic.

A single cycle of the waveform is divided into 128 (08h) small blocks of time. The length of one block is delaytime, which can be set to any value between 1 and 256 (ffh). First the output is made high for a given number of blocks, set by hitime. Then it is made low for a given number of blocks set by lotime. Hitime is typed in and lotime is calculated by subtracting hitime from 128. The output is made low for lotime.

The single waveform cycle is repeated a number of times (lenburst) to produce a burst of pulses at the output to drive the motor for a given length of time. The programming could be adapted to power the motor for a longer time or until a given event occurs.

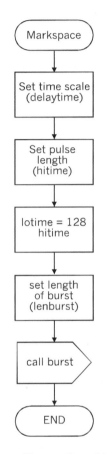

The calling routine of the *Markspace* variable speed program.

The period of the waveform is constant (hitime + lotime) but the ratio hitime:lotime can be varied over the range 128:0 (motor fully on) to 0:128 (motor off). In practice, the motor requires a minimum hitime in order to keep it turning slowly so the full ratio range can not be used. Similarly with LEDs.

The listing overleaf is complete except for the PIC type, configuration and equates. The motor will be controlled through channel RC0, using a transistor switch. The flowchart of the burst subroutine is on p. 256.

```
          lenburst equ 23h
          hitime      equ 24h
          lotime      equ 25h
          hicount     equ 26h
          locount     equ 27h
          delayx      equ 28h
          lencount    equ 29h
          delaytime   equ 2ah

          org 00h
          goto start
          org 04h
          goto start

start
          bsf status, 5      ; Bank1.
          clrf trisc         ; Port C all outputs.
          bcf status, 5      ; Bank2.
          bsf status, 6
          clrf ansel         ; Digital input/output.
          clrf anselh        ; Digital input/output.
          bcf status, 6      ; Bank0.
          bcf status, 5

; Program begins here
          clrf portc
          movlw 010h         ; Set length of unit time-block.
          movwf delaytime
          movlw 70h          ; Set time-blocks for high output.
          movwf hitime
          sublw 080h         ; Calculate low time-blocks.
          movwf lotime
          movlw 050h         ; Set length of burst.
          movwf lenburst
          call burst         ; To produce burst of pulses.
          bcf portc, 0
          bsf portc, 3       ; LED on RC3 signals 2nd burst on.
          movlw 02h          ; Set parameter for 2nd burst.
          movwf hitime
          sublw 080h
          movwf lotime
          movlw 0C0h
```

```
        movwf lenburst
        call burst
        bcf portc, 3


endit    goto endit

; Subroutines

burst
        movf lenburst, w      ; lencount = lenburst.
        movwf lencount
speed
        movf hitime, w        ; hicount = hitime.
        movwf hicount
        movf lotime, w        ; locount = lotime
        movwf locount
        bsf portc, 0          ; Turn output on (high)

hiphase
        call delayit          ; Remains high while hicount
        decfsz hicount, f     ;   is counted down to zero.
        goto hiphase
        bcf portc, 0          ; Turn output off (low).
lophase
        call delayit          ; Remains low while locount
        decfsz locount, f     ;   is counted down to zero.
        goto lophase
        decfsz lencount, f    ; Burst continues while
        goto speed            ;   lencount is counted dowm
        return                ;   to zero.

delayit
        movf delaytime, w     ; Produces a delay while
        movwf delayx          ;   delayx (= delaytime)
nextloop                      ;   is counted down to zero.
        decfsz delayx, f
        goto nextloop
        return

    end
```

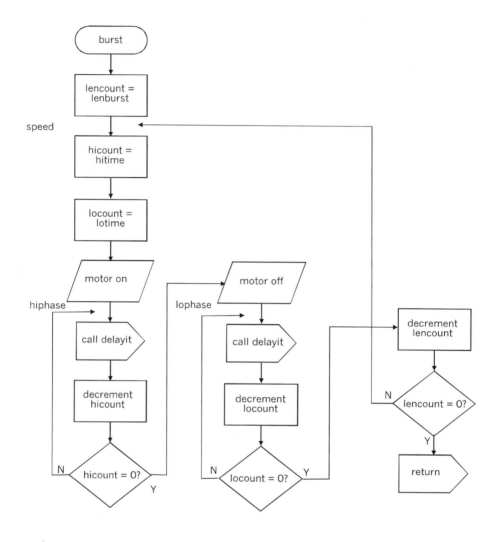

The *burst* subroutine produces a train of pulses of fixed frequency and length
but with variable mark:space ratio.

The routine can be used to give a more realistic action to a mobile robot. Instead of
starting and stopping with a jerk the robot accelerates smoothly from rest, and gracefully
decelerates to a halt. The *burst* subroutine is called with fixed values of delaytime and
lenburst and an initial value for hitime. It then runs in a loop in which hitime is
gradually increased or decreased.

Machine-gun sound effect

This routine generates a burst of white noise which sounds reasonably like a burst of machine-gun fire. It makes a good addition to the repertoire of a military toy robot. The sound is variable so it could have other applications.

The random number generator (pp. 161-163) is made to run in a loop, generating a sequence of values of *randval* at high speed. As each new number is generated, its bit<7> is read (by ANDing *randval* with 080h,) and channel RC7 of port C is set to the same value.

The result is that RC7 outputs a random sequence of values. The output goes to a transistor switch with a speaker in its collector circuit (p. 89). The random levels produce a burst of white noise.

The routine is a short one and calls on *delay*, *shortdelay*, and *randno*:

```
        bsf randval, 0          ; Seeding random numbers.
        movlw 050h              ; Length of burst.
        movwf count
repeat
        call randno
        movlw 080h
        andwf randval, w        ; Get bit <7>
        movwf portc             ; Send it to RC7.
        call shortdelay
        decfsz count, f
        goto repeat
        bcf portc, 7            ; In case last value is 1.

endit   goto endit
```

With count set to 050h, the sound burst is about one second long. If the burst is made longer it develops a regular beat which spoils the effect. This happens because the sequence is not truly random and repeats after 127 calls to *randno*. It would be possible to extend the virtual shift register to, say, 16 bits and give bitn and bitm the values of bits <15> and <14> respectively. The sequence then repeats after every 32767 calls.

6.4 The Quester

This is a mobile platform with room to install as many sensors and actuators as the 16F690 can handle. It is ideal for experimenting with new ideas in the fields of robot mechanics, robot electronics and robot programming.

There is room on the platform for several sensors and the platform can readily be enlarged to take more. The modular design of the electronics give flexibility and the scope for extension. With a PIC at the heart of it all, the *Quester* is ready to explore new types of robotic action.

Specification

Two decks construction:
> Lower: motors, wheels.
> Upper: PIC, sensors, actuators.

Two batteries: 12 V (motors), 6 V (logic).
Three wheels: 2 drive, 1 castor.
PIC 16F690.
Sensors: switches, bumpers, light, IR.
Actuators: motors, LEDs, buzzer/siren.
Control panel on upper deck.

Programs:
> Program selector
> Wanderer (avoids obstacles)
> Light seeker
> Line follower
> Prisoner (can not cross a line)
> Maze runner (on the Web)

Mechanics

The decks are cut from sheet plastic (we prefer 3 mm expanded PVC), plywood, or hardboard. The drawing opposite shows the lower deck. First cut a square, 160 mm × 160 mm. Cut away the recesses for the drive wheels, and cut away the triangular areas at the rear corners. You may need larger recesses if the wheels are greater than 70 mm diameter.

The motors and gears are mounted below the lower deck (see photo opposite). The drive wheels must therefore be large enough to give clearance between the motors and the ground. The photo also shows the base of the castor. When planning the layout of the deck, check that the castor will be able to turn fully through 360° without touching the rear ends of the motors or their supports.

The drawing indicates the locations of the axles and motors. However, the exact positioning depends on the sizes of the motors and other parts.

In our *Quester* the motors operate on 12 V DC, are reversible (important!), and have built-in gearboxes. These reduce the speed of rotation to 70 r.p.m. With 70 mm diameter drive wheels, this gives the robot a speed of:

$$70 \times \Omega \times 70 =$$
$$15 \text{ m/min, or } 256 \text{ mm/s.}$$

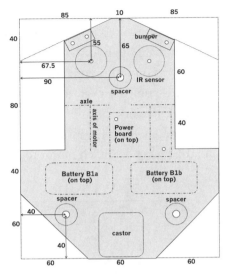

The lower deck seen from below.

This is rather too fast for convenience, so we added external gears (from Meccano) to reduce the speed further. With Meccano gears, use a needle file to re-shape the 'rounded triangular' hole in the 10-tooth cog wheel to fit the motor output shaft. The cog wheels mesh with 50-tooth crown gears.

The lower deck, viewed from below.

These give a further 1:5 reduction in speed, bringing the maximum speed of the robot down to about 50 mm/s, which is much more manageable.

The bearings for the axles consist of two double-right-angled strips (one for each motor shaft). Mount the drive assemblies (motors, gears and bearings and wheels) so that the deck is level when standing on the drive wheels and castor.

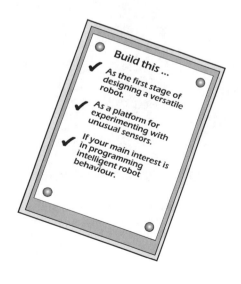

Build this ...

✓ As the first stage of designing a versatile robot.

✓ As a platform for experimenting with unusual sensors.

✓ If your main interest is in programming intelligent robot behaviour.

Most motors have threaded holes for bolts, or have fixing lugs. Others are without any such features and are intended to be mounted in a spring clip. If a clip of the right size is not obtainable, a mounting can be improvised from PVC sheet. Each motor is supported at one end by the angle bracket through which the output shaft passes. The other end is suported in a framework (drawing below) made from strips of PVC board, bolted together and fixed to the deck by an angle bracket. Wind a single layer of PVC insulating tape round the rear end of each motor to help the framework grip the motor.

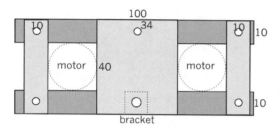

The rear ends of the motors are supported by a framework.

Drill all the holes in the deck before bolting anything to it. Most holes are either 3 mm or 4 mm in diameter, but the three holes for the deck spacers might need to be larger. These are shown in the drawing on p. 269.

Other holes, not in the drawing, are for the brackets supporting the axles and the motors, for the bracket that attaches the framework, and for the castor. Two holes are needed for mounting the power control board, and two more for each hinge of the bumpers.

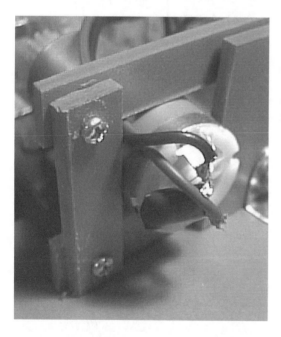

The rear ends of the motor (with smoothing capacitor soldered to its terminals) is supported in a framework. Part of the angle bracket that fixes the framework to the deck is seen on the right.

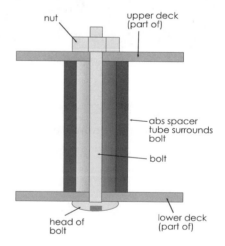

The decks are clamped together by three long bolts (one shown here). Each bolt is surrounded by a spacer tube cut from garden sprinkler tubing. The nut is tightened to hold the decks firmly against the ends of the tube.

A castor, a light-duty furniture type, can be purchased from a DIY store. Try to get one that, when bolted to the deck, supports the rear end of the deck horizontally. Otherwise it is necessary to adjust the mounting of either the castor or the drive wheel assemblies to make the deck level.

The castor is the right size to support the lower deck horizontally.

This view shows the rear end of the right motor supported in the framework. The power control board is visible, mounted on the top surface of the lower deck.

The upper deck (below) is cut from a 160 mm square of expanded PVC board. It has the same overall dimensions as the lower deck. It is mounted on the lower deck so that its rear edge is exactly above the rear edge of the lower deck. There are three holes for the spacer bolts; these must be in exactly the same positions as those in the lower deck. The spacing between the two decks must be large enough to allow access for hands and small tools. However, if the spacing is too large the robot is more likely to topple over on a rough surface. Spacing of 100 mm is about right.

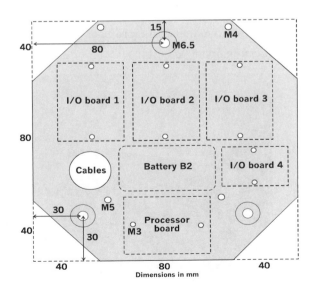

The upper deck, showing the layout of the circuit boards and the positions of the three spacers.

There are several other ways of mounting the upper deck. One way is to cut the spacers from wood dowelling about 20 mm in diameter or 20 mm square timber and fasten the decks by passing screws through holes in the deck and into the ends of the dowelling. Brackets of strip aluminium bolted to the decks are another suggestion.

There is a large hole (about 25 mm diameter) in the upper deck for the cables that connect the two decks. These include the 4-way lead from the processor board to the motor control board. The hole is large enough to pass the sockets that terminate both ends of this cable. Alternatively, you may decide to route this cable another way, in which case the hole could be smaller.

The figure shows the locations of the circuit boards and the battery that supplies the logic circuits. The boards are each to be fixed to the deck by a pair of nylon bolts.

For each board that is to be included in the system, drill a pair of holes to match the holes in the circuit boards. Refer to the stripboard layout figures on pp. 265-275 to find the spacing of these holes.

Note that there are no holes in either deck for fixing the batteries. They are attached with self-adhesive Velcro™ strip. This makes it easy to remove them when replacing or recharging spent cells.

(Above) Three-quarters rear view of the upper deck and control panel (white). The control panel is supported at the front by being hinged to a framework of metal strip. In the foreground are the boards for the bumpers and IR sensors.

(Right) Rear view of the upper deck and control panel. The rear edge of the panel is supported by, but not attached to, two vertical metal strips that are bolted to the deck. The processor board is in the fore-ground and, behind it is the 6 V battery.

The figure shows one of the two vertical supports bolted to the upper deck. These are Meccano pieces. The control panel is hinged to a PVC cross-member joining the upper ends of the supports. At the rear, the control panel rests on another pair of supports, as seen in the photo opposite.

In the photo, the space between the upper deck and the control panel is crammed with circuitry and connecting wire. However, the control panel can be flipped up for easy access. Using wires of several different colours makes it easier to trace connections. The control panel is a rectangle of white polystyrene sheet, but could have been made from the expanded PVC.

The panel carries the switches S1 and S2 for battery B1 (12 V drive motors) and battery B2 (6 V logic) and their indicator LEDs, D1 and D2. The push-button S3 resets the system. The rotary switch S4 selects which of the four programs is to be run. Other switches could be there to activate patterns of behaviour.

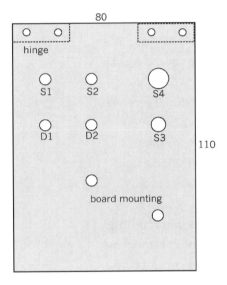

The control panel. There are two holes for mounting the control circuit board.

The completed control panel. You may prefer to label the switches and LEDs.

This is mainly an account of the construction of a particular mobile robot. Many different versions are possible and workable. You may need to substitute other materials and methods of construction, and your eventual design may look very different from the robot shown here. But the same principles apply to all mobile robots. Experiment and innovate!

Electronics

After an overall view, this section looks at *Quester's* electronic system, board by board. A shopping list of the components required is given on pp. 278-279.

The centre of the system is the processor board which carries the PIC16F690 microcontroller. It receives power from B2 through switch S2 on the control panel. LED D2 indicates when the power is on.

There are connections to S3 and S4 on the control panel, to the power control board on the lower deck, and to the input/output boards on the upper deck and elsewhere that interface the sensors and actuators to the PIC.

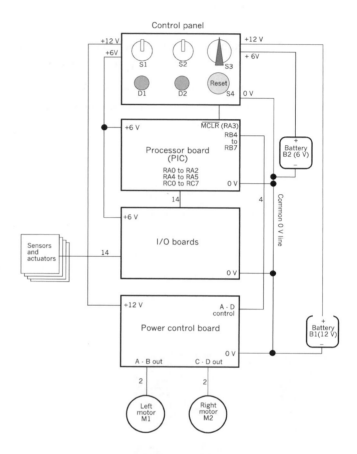

A block diagram of the whole electronic system of the *Quester*.

Controller board

The schematic (opposite) shows that this board has little on it other than the PIC and the circuitry for pulling up the input to channel RA3 (pin 4), which does not have an internal pulll-up. The power switch S2 and the push-button S3 are located on the control panel. As can be seen in the drawings opposite and the photo on p. 266, the main items of the board are a 20-pin socket for the PIC and a large number of PCB pins for I/O.

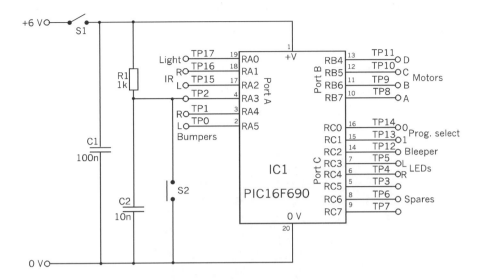

Schematic of the controller board.

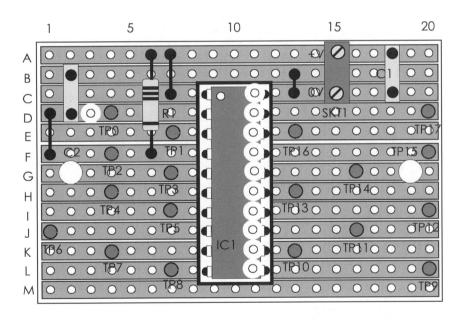

Layout of the controller board. Check the
underneath of the board, using a hand lens
to make sure that all the copper strips are
cut beneath the PIC.

The inter-board connection technique used here provides the greatest possible flexibility when configuring the system for different sets of sensors and actuators.

The diagram shows two M3 mounting holes drilled as G2 and at G19. The board is bolted directly to the upper deck, using M3 nylon bolts and nuts.

A completed controller board. The PIC socket is empty to show the gold insert machined pins. This type of socket is best for the controller IC, which has to be removed for re-programming several times during development and testing.

When the board is completed, the only testing required is for continuity between the individual pin sockets of the IC socket and the corresponding terminal pins.

Separate pins for each I/O connection give flexibility, but you could instead use 2-way or 4-way header plugs for the connections to certain of the I/O boards.

Motor control board

The circuit is the conventional H-bridge. The PCB layout is seen opposite. This version of the circuit provides for two motors operating independently. The connection to the PIC, the control input, is a 4-way header plug on the left in the figures. The output to the two motors is another 4-way header plug, on the right.

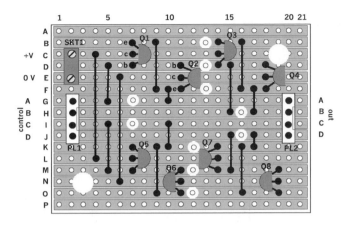

Layout of the motor control board.

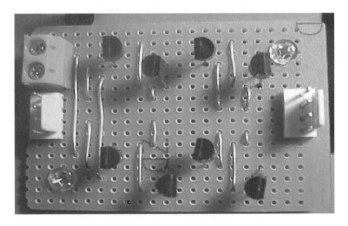

The completed motor control board.

Check the completed board in the usual way by testing continuity and the absence of short circuits. For a functional test, temporarily connect the motors to the 'Out' terminal. Use A and B for the left motor, use C and D for the right motor. The motors should be connected with the same polarity. For example, if A is connected to a given terminal on the left motor, C should be connected to the corresponding terminal on the right motor.

Connect a 12 V battery to SKT1, also connect a pair of flying leads to the battery. If the positive supply is connected to control A and 0 V to control B, the left motor should run in the forward direction. If not, reverse the connections at the motor terminals. Repeat with 0 V to A and positive to B. The motor runs in reverse. Repeat the test for controls C and D.

Control panel

The control panel switches power supplies for the motors (12 V) and the logic (6 V).

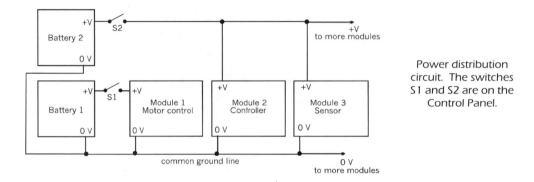

Power distribution circuit. The switches S1 and S2 are on the Control Panel.

Note the 0 V line which connects the 0 V terminal of both batteries to the 0 V terminals of the processor board and all the sensor and actuator boards. In this way all modules in the system have reference to the same ground (0 V) level. Signals may be exchanged between the modules.

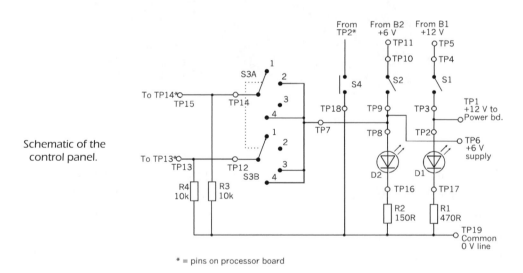

Schematic of the control panel.

* = pins on processor board

The Control panel carries several independent circuits:

- Power switches, S1 and S2, with their indicator LEDs, D1 and D2.

- Program Select rotary switch, S3, with output to the controller board.

- Push-button, S4, with output to the controller board.

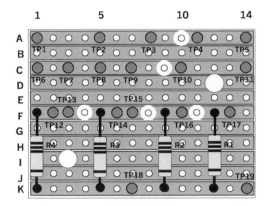
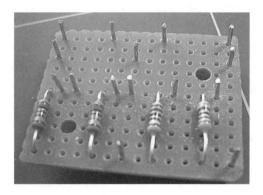

Most of the wiring of the control panel is between the components themselves, already mounted on the panel. There is a small stripboard (above) which connects the on-panel components to the outside modules using PCB pins. This board is mounted on the back of the control panel.

The control panel is quickly tested for continuity and short circuits. Also test the functioning of the circuits particularly the action of S4. The output voltages should follow the binary sequence 00, 01, 10, 11 as the switch is rotated clockwise, with pin TP13 on this board being the more significant bit (goes to pin TP13 on the controller board and to pin 15 of the PIC).

A point to be remembered when running wires to off-panel modules is that the wires must be routed close to the hinge end of the panel. Otherwise it will be difficult to raise the panel when testing or servicing the system. We used small self-adhesive plastic cable grips, fixed on the rear of the front panel. These hold the wires close to the hinged edge of the control panel.

Input/Output modules

The *Quester* has five input modules: a light sensor, a pair of I/R sensors, a pair of bumpers. It also has input from S3 and S4 on the Control panel, as already described. Excluding the drive motors, the robot has three output modules: a bleeper, and two high brightness LEDs.

It is not necessary for you to equip your robot with the same selection. To start with, you may build and install only the bumpers. Experiment with these for a while before adding some of the other sensors and actuators. There are more sensor and actuator circuits described in Part 3 that could extend the behaviour of your robot. Several of the suppliers of robot kits also sell sensor and actuator units that can be connected into the *Quester's* system. Investigate some of these.

Light sensor and LEDs

As there was room for only four boards on the upper deck we decided to build these two modules on the same board. The schematics are on p. 75 (plus an op amp comparator as on p. 73) and p. 91 (left). See Shopping List, pp. 278-279, for component values.

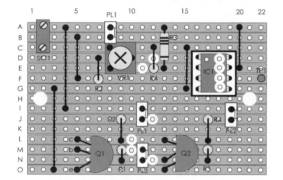

The light sensor occupies the upper half of the board, with the pair of LED switches on the lower half.

The light sensor is nominally an ORP12, but any low-cost LDR can usually be substituted. Preferably it should be about 5 mm in diameter. It has a twin lead soldered to its terminal wires, about 60 mm long and terminating in a 2-way header socket. Polarity is unimportant. The board occupies position 1 (see p. 261) at front right of the upper deck and the LDR is mounted on the front panel (opposite).

The LEDs are mounted by drilling two 1 mm holes 2 mm apart for each. The lead wires are passed through these holes, glue is applied to the base of each LED, and the LED is glued to the panel. The lead wires of each LED are soldered to a pair of wires about 100 mm long, terminating in a 2-way header socket. Polarity is important; the cathode of LED1 should go to the pin at I11 and the anode to J11. The cathode of D2 goes to I19 and its anode to J19. The cathode is usually identified by being the wire lead nearer to the rim on the body of the LED.

The front panel with mounted light sensor and two LEDs. The front panel is PVC cut to fit on the two front supports of the control panel.

To provide directional sensitivity, a 10 mm length of opaque plastic sleeving is pushed on to the LDR. It should be a fairly tight fit. The sleeving is then pushed into a hole in the panel, and this too should be a tight fit.

Note the strip of PVC bolted on top of the hinges. This is for neatness, to cover the gap between the front edge of the control panel and the upper edge of the front panel.

The sensor is tested by connecting 6 V to SKT1, and a voltmeter to TP1 and 0 V. Adjust the setting of VR1 until the output is low (less than 1 V) when the LDR is covered and is high (more than 3.5 V) when the LDR (in its tube) is exposed to light from a daylight window or mains table lamp. It will probably be necessary to readjust the setting when the robot is being tested, depending on ambient lighting conditions.

With the board power supply still connected and the LEDs connected to PL1 and PL2 the LEDs are tested by applying 6 V to the pin at N11; D1 should come on. Similarly, test the switching of D2 by applying 6 V to the pin at M11.

Infrared sensor board

The infrared sensors in this robot are mounted below the lower deck, just behind each bumper, and directed downward. The rims of the shields are about 20 mm above ground level. The probes are built on small squares of stripboard (overleaf). The IR LED (D1) is a 5 mm type able to pass a maximum current of 50 mA. The IR photodiode (D2) is a BP104, which fits conveniently close to the circuit board, but other types can be used. When soldering in the diodes, note that they are mounted with opposite polarity.

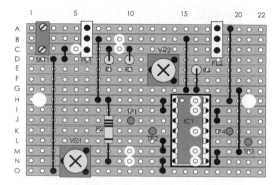

The IR sensor board interfaces the two sensors to the system. The board has two header plugs: PL1 connects to the left sensor and PL2 to the right. Connections are: strip A, +6 V; strip B, output from photodiode cathode; strip C, from IR LED anode. Output is normally taken from the board at TP1 (left) and TP3 (right).

The IR sensor board. The schematic is on p. 77. Both the light sensor (p. 280) and this IR sensor have a comparator built from an IC on the board. We could instead use the PIC's built-in comparator module or the AD converter, but triggering levels would then have to be set by programming instead of by screwdriver.

Underneath view of the robot, showing the two IR sensors. The circuit boards can be seen surrounded by the white plastic shields.

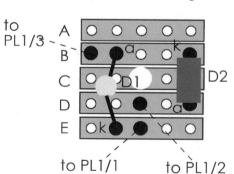

(Left) Stripboard layout of an IR probe. There is a central hole for the supporting bolt.

(Right) The leads of D1 are cut short so that it lies close to the probe board.

The probe boards are enclosed in downwardly directed shields (drawing below). We used PVC pipe caps obtained from the plumbing department of the local hardware store. A 30 mm M3 bolt passes through the circuit board, through a central hole drilled in the shield and through the lower deck. These items are secured by nuts. A second hole is drilled in the shield to one side of centre. This is for the connecting wires (not shown in the figure), which pass through the circuit board *from the rear* and are then soldered.

The mounted probes are seen in the lower photo opposite. They are mounted with their centres about 45 mm apart.

Sectional view of a probe. The inside of the shield may be painted matt black to give greater directional sensitivity, but it was found that this made little difference.

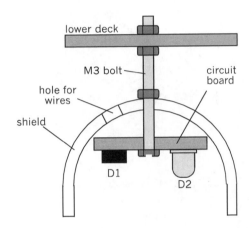

The sensors are tested by using a piece of white card, half of which is painted black. Connect the 6 V power supply and plug in one of the probes. Connect a voltmeter to the output pin. Adjust the preset resistor until the output voltage is low (close to 0 V) when the white area of the card is held 20 mm from the sensor, but rises close to the positive supply voltage when the black area is held there.

Bumpers

These are two rectangular PVC panels, left and right, hinged at the front of the lower deck. They hang vertically but, when pressed, they swing back and contact microswitches. The bumpers need to be able to recover quickly when pressure is removed, so the hinges should be quite loose. Try searching through the stock at the store to pick out the two loosest hinges they have.

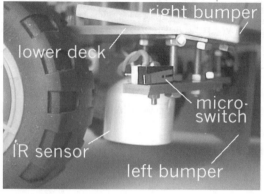

(Right) The microswitch is bolted to a small PVC panel suspended below the lower deck on two bolts. Here the right bumper is flipped back.

The robot encounters an obstacle. It is sensing that its right bumper is pressing against something, closing the right microswitch. Conventionally, bumpers are located at the front of the robot, but it could be useful to have bumpers at the back too, for detecting obstacles while reversing. A bumper, or something similar can also be mounted on one side of the robot for use in wall-following routines.

The bumper board, mounted in position 4 on the upper deck.

The bumper board. Only a small board is needed for bumpers. The resistors are 10 kΩ.

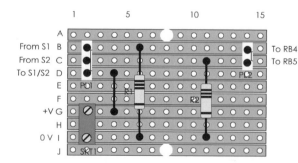

Bleeper board

The board for the bleeper or other audible warning device (AWD) can be mounted in the least accessible position on the upper deck, because there is nothing on the board that will need adjusting. The stripboard layout is adaptable to devices of various shapes and sizes.

For a bleeper, we used a low-cost mini piezo buzzer. It operates on 1.5 V to 28 V, producing a single tone, frequency 3.5 kHz at about 80 dB.

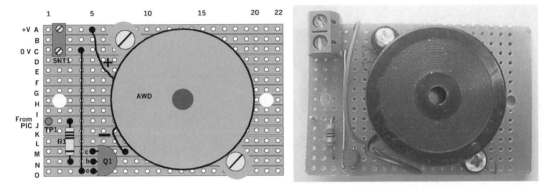

The Bleeper board. For maximum loudness, make sure that the AWD is fixed firmly to the circuit board and the board is fixed firmly to the deck.

Test the board by connecting the 6 V supply. Then apply +6 V to pin TP1 to hear the bleeper sound.

By now we are ready to start making the inter-board connections. First to be made are the power connections on the upper deck. Note the power lines (top left) to the control panel and the common ground line (lower left) to the lower deck. The 6V battery will be fixed to the Velcro patch.

Connections between boards

See the drawings on p. 264 and p. 268, and the table below. The photo on the previous page illustrates the way the 0 V and +6 V lines are daisy-chained from board to board. The PIC is not in its socket, and will not be there until the whole system has been tested.

From board	To	No. of wires	From connector	To connector
Control panel	Processor MCLR	1	Pin*	Pin
Control panel	Processor Program select	2	Soldered to S4	Pins
Light sensor	LDR	2	2-way skt†	Soldered to LDR
Light sensor	Processor	1	Pin	Pin
LEDs	Processor	2 + 2	2-way skt each	Soldered to LEDs
IR sensor	IR sensors	3 + 3	3-way skt† each	Soldered
IR sensor	Processor	2	Pins	Pins
Bumpers	Bumpers	3	3-way skt	Soldered
Bumpers	Processor	2	2-way skt	Pins
Bleeper	Processor	1	Pin	Pin

* Pin = 0.9 mm pin on the board, socket on the lead.

† 2-way and 3-way = polarising header plug on the board, socket on the lead.

Begin testing by checking the power supply to each board. The positive line on the boards should be at 6 V, or 4.8 V to 5 V with rechargeable cells. Check the Input/Output responses as detailed on pp. 102-103. This procedure checks what happens when a given voltage is applied to an output pin (O) in the processor socket, or when stimulating a sensor and measuring the voltage at the input pin (I).

PIC I/O

For later reference the table below lists I/O connections to the PIC. A photocopy of this table mounted on card is handy to have on the workbench.

Port	Name	IC pin	I/O	Connected to	0 =	1 =
A	RA0	19	I	Light sensor	Dark	Light
	RA1	18	I	I/R sensor (R)	White	Black
	RA2	17	I	I/R sensor (L)	White	Black
	RA3	4	I	Push button	Pressed	Released
	RA4	3	I	Bumper (R)	No contact	Contact
	RA5	2	I	Bumper (L)	No contact	Contact
B	RB4	13	O	Motors D (L)	A0 = forward	
	RB5	12	O	Motors C (L)	50 = reverse	
	RB6	11	O	Motors B (R)	60 = spin right	
	RB7	10	O	Motors A (R)	90 = spin left	
C	RC0	16	I	Program select 0		
	RC1	15	I	Program select 1		
	RC2	14	O	Bleeper	Off	Sound
	RC3	7	O	LED (R)	Off	On
	RC4	6	O	LED (L)	Off	On
	RC5-7	5, 8, 9	Spare			

The entries are grouped by I/O port and list all the channels available on the PIC16F690. In the fourth column, the I or O indicates whether the channel is to be configured as an input or an output.

In Port B, the right-hand column lists the binary values used in assembler code to produce the required motion of the robot. If it does not, reverse the connections between the power control board and the terminals of the motor.

A 4-way program select switch with two input lines is not necessary if the controller runs only one or two programs. Instead, the program select lines could be used for other switched inputs such as microswitches for additional bumpers or for limit switches.

Shopping list — electronic (1)

Controller board:

    R1 resistor 1k.
    C1 polyester capacitor, 100n.
    C2 polyester capacitor, 10n.
    20-way d.i.l. turned pin socket.
    2-way screw terminal.
    0.9 mm PCB terminal pins (17 off).
    Stripboard 13 strips x 20 holes.

Motor control board (double):

    Q1, Q3, Q5, Q7 BC639 npn transistor (4 off).
    Q2, Q4, Q6, Q8 BC640 pnp transistor (4 off).
    PCB plugs 4-way (2 off).
    2-way screw terminal.
    Stripboard 16 strips x 21 holes.

Control panel:

    R1 resistor, 470R.
    R2 resistor, 150R.
    R3, R4 resistors, 10k.
    D1, D2 light-emitting diodes, 5mm.
    S1, S2 miniature toggle switches.
    S3 miniature 3-pole, 4-way rotary switch.
    S4 push-to-make push-button.
    0.9 mm PCB terminal pins (19 off).
    Stripboard 11 strips x 14 holes.

Light sensor board:

    R1 light dependent resistor, ORP12 or similar.
    R2 resistor 470R.
    R3, R4 resistor, 10k (2 off).
    VR1 miniature preset potentiometer (trimpot) 10k.
    IC1 CA3140E CMOS operational amplifier.
    8-pin d.i.l. IC socket.
    2-way screw terminal.
    2-way PCB plug and socket.
    0.9 mm PCB terminal pin.
    Stripboard 15 strips x 22 holes.

LED switches (on light sensor board)

    R1, R3 resistors, 470R.
    R2, R4 resistors, 120R.
    Q1, Q2 npn transistors, BC548 or similar.
    D1 - D2 5 mm light emitting diodes, ultra bright (2 off)

Shopping list — electronic (2)

Infrared sensor board
>R1, R3 resistors, 68R (2 off).
>R2, R4 resistors, 22k (2 off).
>VR1, VR2 miniature preset potentiometers (trimpots) 100k (2 off).
>D1, D3 infrared light-emitting diodes, 5 mm (2 off).
>D2, D4 infrared photodiodes, BP104 or similar (2 off).
>IC1 CMOS 4011 quadruple NAND gate.
>2-way screw terminal.
>3-way PCB plugs and sockets (2 off).
>14-pin d.i.l. IC socket.
>0.9 mm PCB terminal pins (4 off).
>Stripboard, 15 strips x 22 holes.
>Materials for the shields.

Bumper board
>R1, R2 resistors, 10k ((2 off).
>2-way screw terminal.
>2-way PCB plug and socket.
>3-way PCB plug and socket.
>MS1, MS2 miniature microswitches, SPST or SPDT (2 off).
>Stripboard 10 strips x 15 holes.

Bleeper board
>R1 resistor, 470R.
>Q1 npn transistor BC548 or similar.
>AWD solid state buzzer or siren, 6 V operating.
>2-way screw terminal.
>0.9 mm PCB terminal pin.
>Stripboard 15 strips x 22 holes.
>Bolts for fixing AWD (2 off).

Miscellaneous
>Connecting wire.
>Solder.
>Bolts for mounting circuit boards.

Programming

It is assumed in this section that you are using the PICkit 2 programmer and software. If you are using another programmer, your listing will be largely the same but with minor differences.

The first lines in any code define variables, as shown opposite, but before these we have a header (optional), state the processor used, and define its configuration.

The listing is written for the 16F690. The configuration wod is 0x33c4 (see pp. 118-119). Note that the directive __config begins with *two* underline characters. The listing continues with directives giving the addresses of registers, the code values of 'w' (working register) and 'f' (the current file register), and the labels of the delay subroutines. There are optional labels for variables used in the modes 2 and 4. You can leave these out if you are not intending to run these modes.

Below is the header, followed by PIC type and configuration word:

```
; * * * * * * * * * * * * * * * * * * * * * * * * * * * * * * * * * * * * * * * * * * * * * * * * * * * * * *
;       ROBOT BUILDER'S COOKBOOK
;
;       Filename: Robot105
;
;       Operates in one of four selectable modes:
;           1) Wanderer - uses bumpers to avoid obstacles
;           2) Light seeker
;           3) Line follower
;           4) Prisoner - wanders randomly within an
;               area enclosed by a line
;
; * * * * * * * * * * * * * * * * * * * * * * * * * * * * * * * * * * * * * * * * * * * * * * * * * * * * * *


        list p=16F690

        __config 0x33c4
```

Next come the equates followed by initialising instructions:

```
status        equ 03h
porta         equ 05h
portb         equ 06h
portc         equ 07h
trisa         equ 05h
trisb         equ 06h
trisc         equ 07h
ansel         equ 1eh
anselh        equ 1fh

w             equ 00h
f             equ 01h
z             equ 02h

delay0        equ 020h
delay1        equ 021h
delayn        equ 022h

;Directive for Mode 2
scans         equ 023h

;Directives for Mode 4
randval       equ 024h
bitn          equ 025h
bitm          equ 026h

;Program begins here

goto start
     org H'0004'
     goto start

start
     bsf status, 5      ;Page 1.
     clrf trisb         ;All port B as outputs.
     movlw H'03'        ;RC0 and RC1 as inputs.
     movwf trisc
     bcf status, 5      ;Back to page 0.
```

```
bsf status, 6        ;Page 2.
clrf ansel           ;Digital I/O.
clrf anselh          ;Digital I/O.
bcf status, 6        ;Back to page 0.
clrf porta           ;Clear all ports.
clrf portb
clrf portc
```

Port A is an input port by default, but Port B must be configured as an output port. To read the program select switch (S3) bits RC0 and RC1 of Port C must be inputs. The rest are to be outputs.

By default, Port C accepts analogue input or output. We need it to be operated only as a digital port. In the 16F690 this is done by setting the bits in the 'analogue select' registers (ANSEL and ANSELH) to 0. Two 'clrf' instructions do this. Clearing all three registers completes the initialising routine.

If you are intending to implement only one of the modes, go ahead and start typing its listing. If you are intending eventually to run all four modes, you need the routine (below) which reads input from the program select switch on the control panel and directs the PIC to the program of your choice. The flowchart opposite shows how this operates.

```
btfsc portc, 1                  ;Test mode select bit 1.
    goto bit1hi
    btfsc portc, 0              ;Test mode select bit 0.
    goto mode2                  ;Light seeker mode.
    goto mode1                  ;Wanderer mode.

bit1hi
    btfsc portc, 0              ;Test mode select bit 0.
    goto mode4                  ;Prisoner mode.
    goto mode3                  ;Line follower mode.
```

Next, the setting of switch S4 is tested by reading RC1. If it is low (switch set to 0X) the program continues by testing bit 0. Depending on the value of bit 0 the program takes us to Mode 1 (00) or Mode 2 (01). A high value of bit 1 (switch set to 1X) causes a branch to the bit1hi label where a test of bit 0 routes either to Mode 3 (10) or Mode 4 (11).

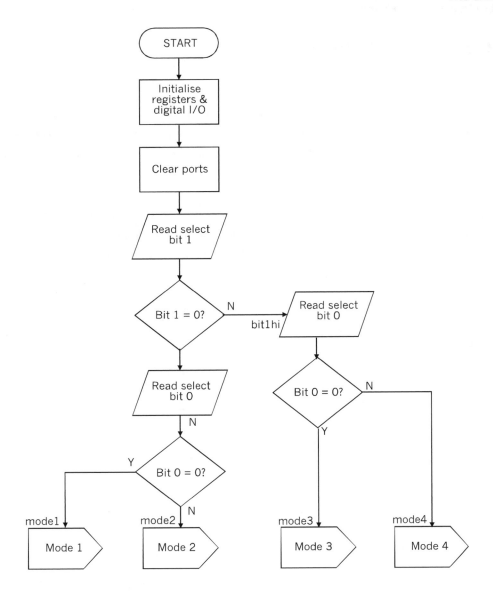

Flowchart of the program
select routine.

If you are intending to install two or more modes, but want to type and test one at a time, it is safer to put in short routines to catch the processor when it attempts to run on from the end of a routine earlier in the listing. The simplest is:

```
mode2      goto mode2
```

An alternative is:

```
mode2      goto $
```

Remember that a listing must always end with the 'end' directive.

Mode 1 — Wanderer

In this mode, the robot bleeps once, then runs across the floor, bumping into furniture, walls and obstacles. Every time it bumps it switches on the LED on the side that it bumped, reverses a short distance, turns away from the side it bumped, switches off the LED and then continues on its way, flashing the LEDs and emitting bleeps as it does so. It repeats this sequence indefinitely.

The *Quester* needs the two bumpers for this mode, but no other sensors. It is a good program for your first on-the-job testing of the robot. Ideally the robot also needs the two LEDs and the AWD. However, these are optional and the program works without them — it is just less noisy and colourful.

The flowchart (right) explains how it works. The listing opposite gives the details. This mode also requires four subroutines, which are listed on p. 286. The *avoidright* and *avoidleft* subroutines are used only in mode1, but could be useful in other programs. The *delay* and *longdelay* subroutines are used by all modes, so be sure to type it in.

The main routine (right and opposite) is a loop that begins by reading the bumpers. If one of these is pressed, the appropriate subroutine is called and the robot takes the avoiding action described earlier.

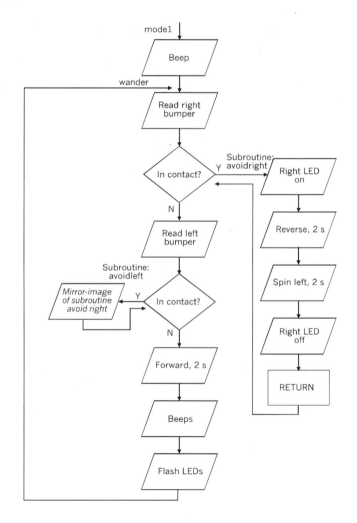

(Right) Flowchart of mode1.

If nothing is detected, the robot runs forward for about 2 s, then stops, bleeps and flashes its LEDs before returning to the beginning of the loop.

The *delay* subroutine is called several times in the loop to provide the necessary gap between switching on the LEDs or AWD and off again. This produces a delay of fixed length, about 0.2 s.

The *longdelay* subroutine is more flexible. Before calling it, a hex value is loaded into the working register. *Longdelay* decrements this repeatedly until it becomes zero. At each stage, *longdelay* calls *delay*. So *longdelay* can produce a delay of any length up to about 50 s in steps of 0.2 s.

Here is the listing of the main loop of Mode 1:

```
mode1
     bsf portc, 2        ;Beep, 0.2 s.
     call delay
     bcf portc, 2

wander
     btfsc porta, 4      ;Read right bumper.
     call avoidright
     btfsc porta, 5      ;Read left bumper.
     call avoidleft
     movlw H'A0'         ;Code for 'Forward'
     movwf portb         ;Both motors forward.
     movlw H'0A'         ;Delay 10 x 0.2 = 2 s.
     call longdelay
     clrf portb          ;Stop.

     bsf portc, 2        ;Short bleep.
     call delay
     bcf portc, 2
     call delay
     bsf portc, 2        ;Long bleep.
     movlw H'08'
     call longdelay
     bcf portc, 2
     bsf portc, 3        ;Flash both LEDs.
     bsf portc, 4
     call delay
     clrf portc
     goto wander         ;Repeats without end.
```

These are the subroutines:

```
delay
    decfsz delay0, f
    goto delay
    decfsz delay1, f
    goto delay
    return
longdelay
    movwf delayn
repeat
    call delay
    decfsz delayn, f
    goto repeat
    return
avoidright
    bsf portc, 3        ;Right LED on.
    movlw 050h          ;Both motors reverse.
    movwf portb
    movlw 0ah           ;Delay = 10 x 0.2 = 2 s.
    call longdelay
    movlw 090h          ;Spin left.
    movwf portb
    movlw 0ah
    call longdelay
    clrf portb          ;Stop.
    clrf portc          ;LED off.
    return
avoidleft
    bsf portc, 4        ;Left LED on.
    movlw 050h
    movwf portb
    movlw 0ah
    call longdelay
    movlw 060h          ;Spin right
    movwf portb
    movlw 0ah
    call longdelay
    clrf portb
    clrf portc
    return

end
```

In this program the inputs from the bumpers are polled at frequent intervals. In the 16F690, Port A inputs can be configured to cause an interrupt on a change of input. We could instead use an interrupt routine to call the avoding action. Details are in the PIC Data Sheet. However, this makes the program more complicated and, since it is no problem if the robot proceeds in short steps, it is easier to use a polling routine.

Mode 2 — Light seeker

This program is best run with the robot in a curtained or low-lit room. It works most effectively if there is only one source of light. There should not be any obstacles on the floor. The robot needs the light sensor and the pair of bumpers.

The *Quester* is made to detect the direction of the source by spinning round while continually reading the input from the light sensor. The tube on the light sensor restricts the angle of view to about 10 degrees ahead. As soon as the sensor detects light ahead, the robot stops spinning, enters the *advance* routine, and moves forward for about 3 s. Then it stops and reads the input from its bumpers.

If the bumpers signal that they have made contact, it is probable that the robot has reached the lamp or window. It reverses a short distance turns off its LED and the program ends. If the bumpers are not in contact, the light sensor is read again and, if light is still seen, it returns to its *advance* behaviour. If no light is visible now, it goes back to the beginning and starts to scan for light again.

The total number of scans is limited to 128, where a scan is a single 0.2 s spin to the left. It may take several scans to find the light again each time it loses it. This value is stored in the register scan at the start of the program. Whenever the robot loses track of the light scan may be decremented several times. If, after 128 scans, light has not been detected, it may be that a bulky object is blocking the light, or the lamp has been switched off. The robot goes to the *distress* routine, in which it turns off its LEDs and remains stationary, bleeping for help.

Try to extend the searching activity by amending the *distress* subroutine. Instead of stopping and bleeping, the robot could be programmed to set off for a new location. From this location it scans for light again. It may be able to see it again from its new location. It repeats this several times until it finds the light again.

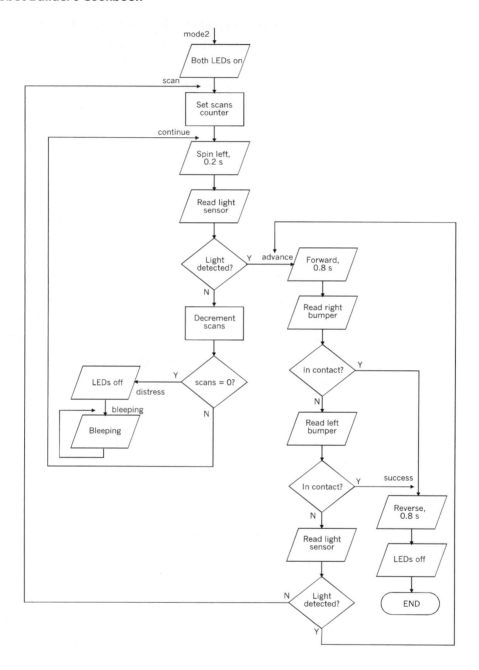

Flowchart of mode2.

Note that this mode calls on the *delay* and *longdelay* subroutines (p. 286). If they are not already included, add them to the listing, shown opposite and on p. 290.

```
mode2
     bsf portc, 3          ;Right LED on.
     bsf portc, 4          ;Left LED on.

scan
     movlw 80h             ;Max. 128 scans
     movwf scans

continue
     movlw 90h             ;Spin left.
     movwf portb
     call delay
     clrf portb            ;Stop.
     bsf portc, 2          ;Bleeper.
     call delay
     bcf portc, 2
     btfsc porta, 0        ;Light detected ahead?
     goto advance          ;Yes  - advance toward light.
     decfsz scans, f       ;Counting down number of scans.

distress
     clrf portc            ;LEDs off.
bleeping
     bsf portc, 2          ;Bleeping.
     call delay
     clrf portc
     movlw 10h
     call longdelay
     goto bleeping

advance
     movlw a0h             ;Forward.
     movwf portb
     movlw 10h
     call longdelay
     clrf portb            ;Stop.
     btfsc porta, 4        ;Right bumper in contact?
     goto success          ;If in contact.
     btfsc porta, 5        ;Left bumper in contact?
     goto success          ;If in contact.
     btfsc porta, 0        ;Light still visible?
     goto advance          ;Yes, continue in same direction.
     goto scan             ;No, look for light.
```

```
success
        movlw H'50'          ; Reverse away from lamp (hot!)
        movwf portb
        movlw H'10'
        call longdelay
        clrf portb           ; Stop.
        clrf portc           ; LEDs off.
        goto $               ; Wait (another way of doing it).
```

Mode 3 — Line follower

The *Quester* follows a line marked on the floor or other surface. For this task, it needs the pair of infrared sensors. The ultra-bright LEDs on the front panel are not essential, but they add to the visual appeal of the robot, especially when it is operating in low light. It can, of course, operate in darkness or in bright room lighting. There is also the practical point that when the robot is spinning left or right to keep on a curving line, the LED on the side to which it is spinning comes on. This helps you to check that the line following is working correctly.

The line is either made from adhesive black strip, from black card, or is painted on the running surface. Before preparing the line, confirm that the line really does absorb infrared. Some black papers and card reflect it, so the black line is indistinguishable from the white or light coloured surface. We found that painting a line on thin white card using Payne's Grey acrylic paint, gave good results. Presumably, Lamp Black acrylic would be just as good.

The line should be about 20 mm wide. Prepare a loop line with straight sections and curves. The radius of each curve should be 70 mm or more. With sharper bends the robot finds itself attempting to cross the line, and stops.

The flowchart and diagram opposite explain how the robot reads the left and right sensors to find out if they are detecting reflected IR. Normally, when the robot is running along a straight line, the sensors are positioned on either side of the line. The situation is as at (a) the top of the diagram. The robot moves forward for 0.2 s and reads the sensors again.

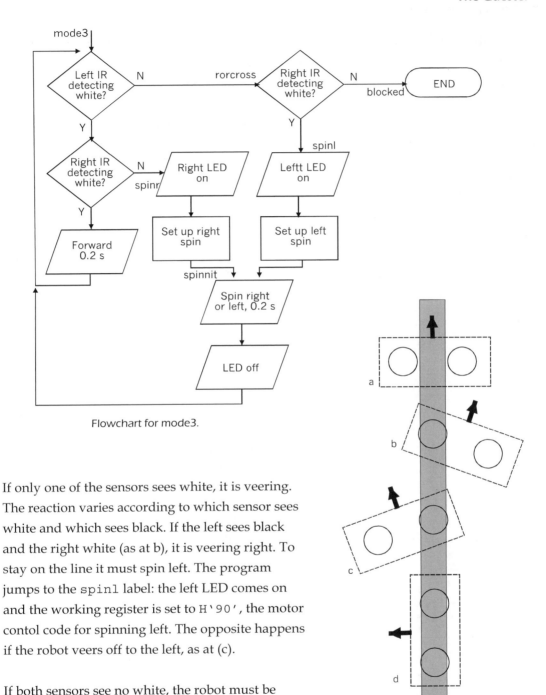

Flowchart for mode3.

If only one of the sensors sees white, it is veering. The reaction varies according to which sensor sees white and which sees black. If the left sees black and the right white (as at b), it is veering right. To stay on the line it must spin left. The program jumps to the spinl label: the left LED comes on and the working register is set to H'90', the motor contol code for spinning left. The opposite happens if the robot veers off to the left, as at (c).

If both sensors see no white, the robot must be crossing the line (as at d). It stops and the program ends. This is where you might program it to do something more exciting, as suggested later.

What the sensors see, in the four possible cases.

In cases (b) or (c) program jumps to *spinl* or *spinr,* depending on which direction it needs to spin. The working register (w) is loaded with the appropriate code and the program goes to *spinnit.* There, it uses the value in w to spin the robot in the required direction.

The program repeats and the robot runs round and round the loop indefinitely, or until it is blocked when attempting to cross the line.

The listing is:

```
mode3
     btfsc porta, 2      ;Left IR sensor.
     goto rorcross       ;Is veering right or crossing.
     btfsc porta, 1      ;Right IR sensor.
     goto spinr          ;Is veering left.
     movlw a0h           ;Forward.
     movwf portb
     call delay
     clrf portb          ;Stop.
     goto mode3          ;Repeat endlessly.
rorcross
     btfsc porta, 1      ;Right sensor.
     goto blocked        ;Is crossing.
     goto spinl          ;Is veering right

spinr
     bsf portc, 3        ;Right LED on.
     movlw 60h           ;Spin right.
     goto spinnit

spinl
     bsf portc, 4        ;Left LED on.
     movlw 90h           ;Spin left.

spinnit
     movwf portb         ;Make it spin.
     call delay
     clrf portb          ;Stop.
     clrf portc          ;LED off.
     goto mode3

blocked
     goto $              ;Wait (or further action).
```

Many amendments and extensions can be made to this program. For example, a simple amendment is to program it to follow a white line on a dark surface.

You can experiment with the lines. If the robot comes to an end of a line, it will continue more or less straight ahead for a short distance. Then it can detect the beginning of another line and follow it. Use this behaviour to make the robot follow a figure-of-eight loop.

The robot can also run a maze of branching lines. At each branch there is a strip across the junction to make it stop. Then it makes a random selection to go either left or right. and arrives at one of several endings. Only one of these is 'home'. Program it to remember which selection it made at each branch and, eventually, to learn to run directly home every time. *Quest05.asm*, a maze runner program on this theme, is available for downloading from the companion website (p. vi).

Mode 4 — Prisoner

This mode requires the pair of IR sensors and, optionally a pair of LEDs and the bleeper.

The *Quester* is placed in an area completely surrounded by a continuous line. It is unable to escape because it is programmed not to cross the line. It wanders randomly and indefinitely within its two-dimensional prison. However, you could leave a narrow gap in the line and see how long it takes to find its way out, and then bleep triumphantly.

The flowchart (overleaf) begins with a simple loop that keeps the robot moving forward with both LEDs on for as long as neither sensor detects the black line. However, if the left sensor or the right sensor detects a line, the program jumps to the *randomspin* routine.

In the *randomspin* routine, it turns off the LEDs and sounds the bleeper while it reverses for 3.2 s. Then a digit (0 or 1) is randomly generated using a version of the random number routine described on pp. 161-163. The number is stored in *randval*. After this, the code to be sent to the motors is first set to 90h (spin left) but is changed to 60h (spin right) if the random digit is 1. After spinning in the randomly-chosen direction and some more flashing and bleeping, the robot resumes its forward progress. The program runs indefinitely.

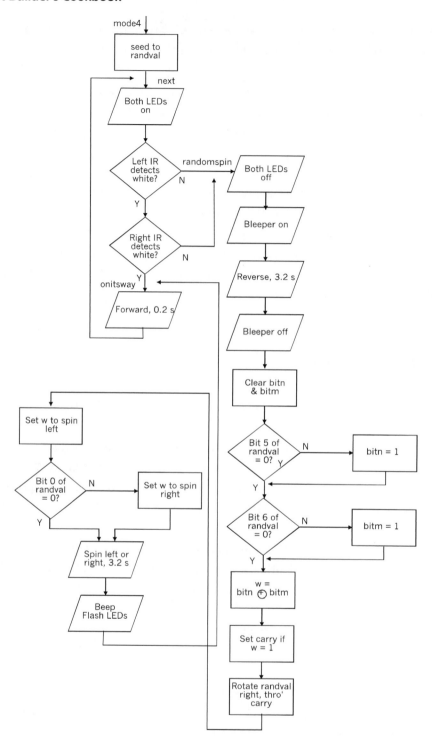

Flowchart for mode4.

When typing in this program, you need to include the three extra equates listed on p. 281. Also include the *delay* and *longdelay* subroutines.

```
mode4
      movlw 33h
      movwf randval           ;Seed value to randval.


next
      bsf portc, 3            ;both LEDs on
      bsf portc, 4
      btfsc porta, 2              ;Left IR sensor.
      goto randomspin        ;Blocked to the left.
      btfsc porta, 1              ;Right IR sensor.
      goto randomspin        ;Blocked to the right.

onitsway
      movlw a0h               ;Forward.
      movwf portb
      call delay
      clrf portb             ;Stop.
      goto next              ;Repeat endlessly.

randomspin
      clrf portc             ;LEDs off.
      bsf portc, 2           ;Bleeper on.
      movlw 50h              ;Reverse.
      movwf portb
      movlw 10h
      call longdelay
      clrf portc             ;Bleeper off.
      clrf portb             ;Stop.

      clrf bitn
      clrf bitm
      btfsc randval, 5       ;Getting n.
      bsf bitn, 0            ;If n = 1.
      btfsc randval, 6       ;Getting m.
      bsf bitm, 0            ;If m = 1.
      movf bitn, w           ;n to w.
      xorwf bitm, w          ;XOR m and n, result in w.
      addlw 0ffh             ;Set carry if w = 1.
```

```
        rlf randval,f          ; New random number in randval
        movlw 90h              ; Set to spin left.
        btfsc randval, 0       ; Get bit 0 of random number.
        movlw 60h              ; Change setting to right.
        movwf portb            ; Spin left or right.
        movlw 10h
        call longdelay
        clrf portb             ; Stop.
        bsf portc, 2           ; Bleep.
        bsf portc, 3           ; Flash LEDs.
        bsf portc, 4
        call delay
        clrf portc
        goto onitsway
```

Developing the *Quester*

There is plenty of scope for combining some of the routines in this specification to make longer and more complex programs — the PIC16F690 still has plenty of program memory to spare.

Consider building some of the other sensors described in Part 3. Add them to this robot. For instance, the sound sensor has several applications ranging from responding to a hand-clap (everyone in the room has to keep very quiet when this is running) to allowing a pair of robots to communicate by using sound signals. The ultrasonic transmitter and receiver provide another way of detecting and locating obstacles.

The *Gantry*'s camera is a very sensitive and directional light sensor, which could be applied to light-seeking or light-avoiding programs.

Too many wires visible? This is a research robot so wires are inevitable. Some people like to see the 'works' but, if you prefer a sleeker look, make a 'skin'. Use PVC sheet, foam board or thick cardboard. The skin drops over the *Quester*, to hide the circuit boards, batteries and wiring,

Gantry robots are often found on the factory floor, but less often on the hobbyist's workbench. A gantry robot usually consists of a main frame that supports a pair of overhead rails, the **y-rails** (see drawing). The y-rails carry the wheeled **y-frame**. This bears a second pair of rails, the **x-rails**, which carry the wheeled **x-frame**.

The tool or load is suspended from the x-frame and may be raised or lowered. By moving the frames in the x- and y-directions and by raising or lowering the tool, the tool may be placed precisely at any point in or above the working area.

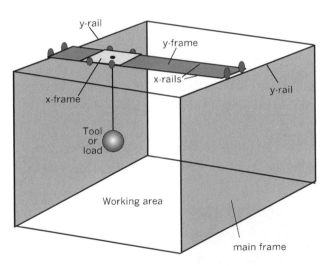

The architecture of a typical gantry robot.

For applications such as playing board games a gantry has the advantage that the location of the tool is very precisely known. This is not usually the case with simple mobile robots. The techniques for finding their location are usually subject to errors, and the errors are cumulative.

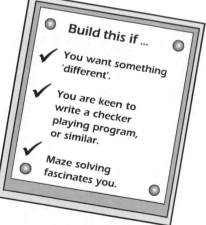

Build this if ...

✔ You want something 'different'.

✔ You are keen to write a checker playing program, or similar.

✔ Maze solving fascinates you.

Mechanics

Specification

Aluminium extrusion construction.
Main frame plus x- and y-frames.
Two motors for x- and y-frames.
One motor for tool, up/down.
PIC16F690.
Two IR or Hall Effect sensors for x- and y-location.
Tools: hook, gripper, paint brush, laser beam, camera.

Programs:
  Moving from A to B
  Moving from C to D
  Scanning
  Operating the hook.
  Operating the gripper*
  Horseshoe game*
  Maze solving*
  * On the companion site.

Aluminium extrusions are ideal for this project. In our prototype we mainly used 15 mm × 15 mm channelling, 3 mm thick. We used 3 mm expanded PVC for the panels on which electronic modules are to be mounted.

Because the design needs to be adapted to the parts available, especially the motors and gearboxes, it is best to start in the middle and work outwards. The central unit of the robot is the x-frame. This is built from two lengths of aluminium channelling, 100 mm long, bolted to two cross-members, 50 mm long.

The x-frame, like most of the *Gantry*, is bolted together using M3 bolts and nuts, with shake-proof washers. Nuts are locked with thread-locking anaerobic adhesive.

Drill holes in the centre of each cross-member for tying the thread used to pull the x-frame along the rails. Drill holes in the longer members for mounting the wheels and the PVC panel, which measures 70 mm × 50 mm.

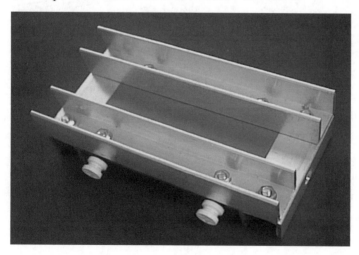

The x-frame consists of four lengths of channelling bolted together. The structure is held 'square' by the PVC panel bolted across the long members. The same bolts are used to secure the wheel base-plates.

The most suitable wheels we found for this robot are Lego™ parts (right). We used the smaller of the two sizes (approx. 8 mm diameter) supplied in a bulk pack of wheels. The wheel hubs (remove the tyres) snap firmly on to the square base-plate, and have a low-friction action ideal for this application. The base-plate has two projections for accepting wheels; saw off the unused projection. Drill a 3 mm hole through the centre of the base-plate for bolting the plate to the longer members of the frame.

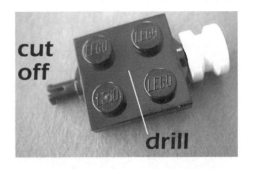

A base plate (black) with one wheel (white). The groove in the wheel is just wide enough for it to ride firmly on a rail 3 mm thick.

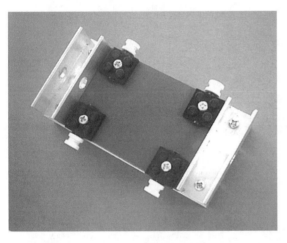

The x-frame seen from below, to show how the four wheels are mounted. At this stage there is no provision for mounting the tool.

When the frame is assembled, the grooves in the opposite pairs of wheel-hubs are 57 mm apart. Measure this distance before proceeding. If the grooves in the wheels are not 57 mm apart, the width of the next stage, the y-frame, will need to be changed to fit.

The dimensions of the x-frame as described above are minimal, to keep the size of the completed *Gantry* within reasonable bounds. However, there is no problem in making it longer or wider (or both) to accommodate certain kinds of tool.

The y-frame (overleaf) is considerably longer than the x-frame because it has to allow the tool to traverse the width of the working area and a little to either side. We will assume that the working area is to be like a chess-board with 64 squares arranged in 8 rows and 8 columns. If each is 30 mm square, the minimum length of the y-frame will be:

(width of cross-member × 2) + (length of motor panel) + (length of x-frame) + (width of working area) + (free space on either side of working area).

In the prototype, the dimensions are $(15 \times 2) + 80 + 100 + (30 \times 8) + 40 = 450$ mm.

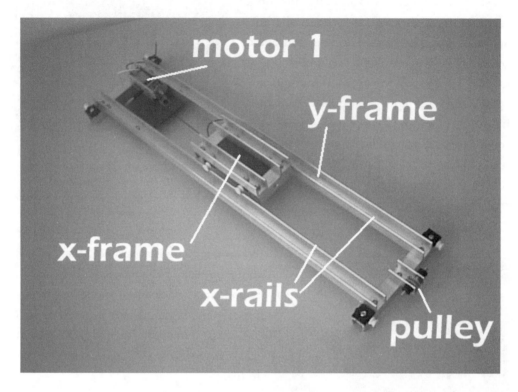

The completed y-frame, showing the x-winch motor mounted on the motor panel at the left-hand end. The pulley for the counterweight cord is at the right-hand end.

As shown on the right, the cross-members project beyond the width of the frame. The base-plates are approximately 16 mm square, so the cross-members need to project 17 mm on each side. The total length of a cross-member is:

(spacing between x-rails) + (width of channelling) × 2 + (projection) × 2

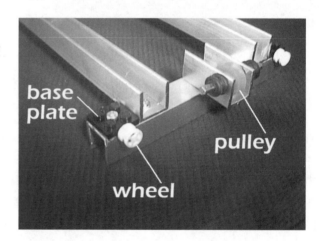

The pulley end of the y-frame. The pulley mounting must project at least 20 mm beyond the end of the frame to allow the cord to clear the y-rail.

In the prototype, the dimensions are:
57 + (15 × 2) + (17 × 2) = 121 mm

The main frame consists of a front rail, a rear rail and two side rails, bolted together at their corners, as shown below. These are cut from aluminium channelling. They are bolted together, the front and rear rails with open sides up, and the two side rails with open sides down. The lengths of the rails are optional but remember that the working area of the robot is less than that enclosed by the frame. The spacing between the side rails must equal the distance apart of the grooves in the opposite wheels of the y-frame. In the prototype, this distance is 455 mm. The front and rear rails are 485 mm long (with 15 mm channeling), and the side rails are 510 mm long.

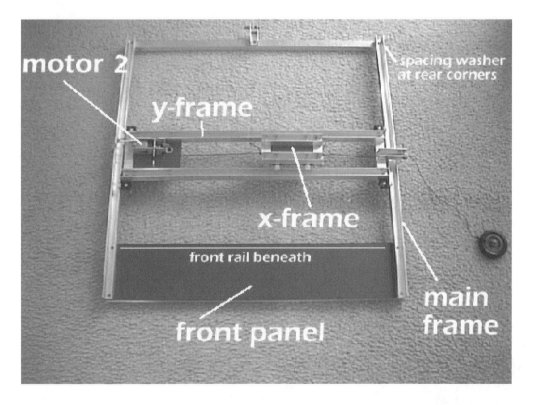

The completed main frame. The y-frame runs from front to rear along the side members of the main frame. The front panel is used for mounting PIC1 and other electronic control circuits.

The frame includes the front panel, 485 mm × 100 mm, in expanded PVC. This has 3 mm holes at its corners and is bolted *between* the front rail and sides rails at the front corners. This means that washers or nuts of equal thickness must be bolted between the rear and side rails at the rear corners.

Note that the other function of the front panel is to keep the frame square.

The *Gantry* is completed by adding four legs, cut from aluminium channelling. Their length is optional and in the protoype they are 340 mm long. The legs must be long enough to allow the counterweights to pull the two frames the full distance along their rails. The front two legs are bolted to the side rails, just clear of the front panel. When drilling the hole in the top end of the leg, drill through *both* sides of the rail, to make an access hole for the screwdriver.

Assembly of the two rear legs is slightly more complicated as they are at the rear corners and have to be bolted to the rear rail and the side rail. When marking the leg for drilling allow for the spacing washer or nut between the rails (see above).

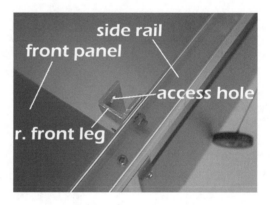

(Right). Close-up to show the front right leg bolted to the right side rail. Note the screwdriver access hole.

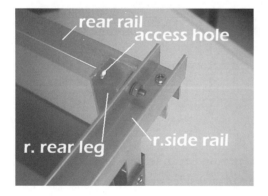

(Left). Close-up to show the right rear leg bolted to the right side rails and the rear rail (bolt not visible) for this rail. Note the screwdriver access hole.

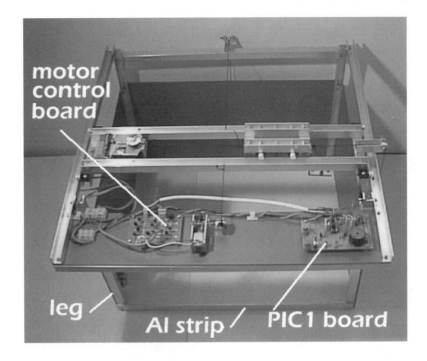

The completed *Gantry* with the winches and some of the
electronic units in position. The aluminium strip joining the front
two legs helps to make the gantry more rigid.

The completed *Gantry* is shown above. The next step is to install the winches and the
microswitch limit switches. First consider the winches — where they are and what they
do. The x-winch moves the x-frame along the x-rails. Drive is provided by a small DC
motor M1, running on 3 V. The winch is connected to the x-frame by a cord which is
wound in or out, depending on the direction of rotation of the motor shaft (below). The
cord is kept taut by a counterweight attached to the other end of the frame. The same
method is used for driving the y-frame.

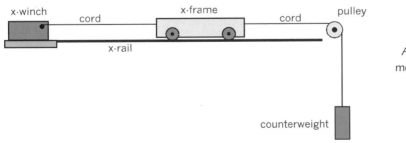

A simple winch-driven
mechanism for moving a
wheeled frame.

The counterweight needs to be heavy enough to pull the x-frame along as the winch unwinds, but not so heavy that the winch can not wind it up. In the prototype, the counterweight is four fishing tackle weights of 35 g. Any compact object of about the same weight will do. The y-frame is moved by a similar mechanism, its winch being mounted on the front panel of the main frame, and with eight 35 g counterweights.

Motors

This project has three DC motors acting as winches. Motor 1 is mounted at the centre of the front panel and moves the y-frame. M2 is mounted at one end of the y-frame and moves the x-frame. M3 is mounted on the y-frame and raises or lowers the tool.

The project may have one or more other motors for driving tools such as a gripper and a paintbrush.

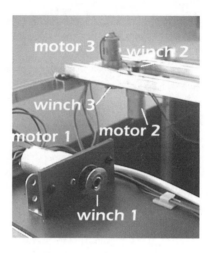

The three winches. Motor 3 is above the panel and its winch is below. Motor 2 is the other way up. Winches 2 and 3 are both close to the central axis of the y-frame. This is to avoid skew forces derailing the x-frame. Winch 1 is at the centre of the front panel.

All three winch motors are powered and controlled from the motor power board, which is mounted on the front panel. The paired leads to M2 and M3 run from a terminal block at the left end of the main frame to a block on the panel of the *y*-frame. The lead must be long enough to allow the y-frame to run freely from front to rear.

Choose the most flexible wire you can get. In a full-sized industrial gantry the leads are not a constraint on the motion of the frames. In a project such as this, the most flexible connection wires commonly stocked by hobby electronics stores are relatively too stiff.

We selected the motor for its small size, built-in gearbox, long protruding shaft, and low cost. The motor has a fixed reduction gearbox. On a 12 V supply, the output shaft rotates at 36 rpm with an output torque of 12 kg cm. The motor is rated to run on any voltage in the range 4.5 V to 18 V, at proportional speeds and torque. We found that the frame and tool moved too slowly if the cord was wound directly on to the shaft.

To increase frame speed, fit metal bobbins on the shafts of the drive motors. The bobbins used in sewing machines are suitable. A motor shaft is usually 2 or 4 mm in diameter but the standard bobbin needs a 6 mm shaft.

Plastic rod cut from the right kind of coat-hanger is a very tight fit in the bobbin. Force a 10 mm length of rod through the central hole of the bobbin. Then drill a 4 mm hole through the rod, along the axis. This must be a tight fit on the shaft of the motor.

The bobbin also has the advantage that it confines the cord close to the central axis of the y-frame when winding on, so preventing skew forces from derailing the x-frame.

Before fitting the bobbin, it may be necessary to cut the output shaft short to save space on the front panel.

The motor/gearbox unit shown in the figure has two M3 threaded holes on either side of the output shaft. Using these, it is bolted to a small PVC panel which is attached to the front panel by two angle brackets.

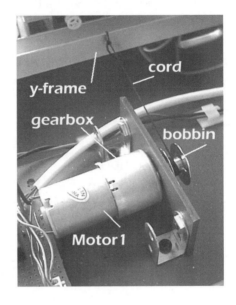

The y-winch mounted in the middle of the front panel, bolted to a small rectangle of PVC board. The motor output shaft has a bobbin on it. Cord runs from this to the y-frame.

A bobbin may not be necessary with a speedier motor. In this case the cord is wound directly around the shaft. It is anchored to the shaft by tying a knot close to one end, threading this through a 5 mm length of 3 mm heatshrink tubing, pushing this on to the shaft and shrinking it until it grips the shaft tightly. The knot prevents the cord from being pulled out.

If the cord is wound directly on the shaft, it is preferable to secure two small discs or wheels to the shaft about 10 mm apart. The cord is wound between these guides.

Limit switches

The x- and y-frames have IR or Hall effect magnetic sensors for determining the x- and y-positions of the tool. In addition, a pair of microswitches mounted on the main frame are used as limit switches. These detect when the x- and y-frames are in their base positions. They reliably put the tool at the front right corner of the frame. From there it uses the IR or Hall effect sensors to find its way to any other place in the working area. In other words, the limit switches provide a starting point to return to if the software loses count of the markers.

The limit switch for the y-direction is microswitch MS1. This is bolted at the top of the front right leg so that the y-frame closes it when it comes to the front limit of its travel.

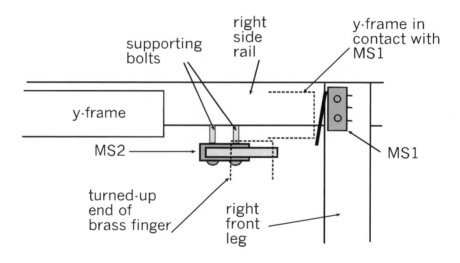

Both MS1 and MS2 are closed when the x-frame is at the front
right corner of the main frame.

MS2 is bolted underneath the right side-rail of the main frame. A finger projects from one end of the x-frame (see opposite). The finger is shaped so that it projects under the end of the y-frame to contact the microswitch on the main frame. Its vertical end presses against the lever of MS2 as the x-frame reaches the right end of its travel.

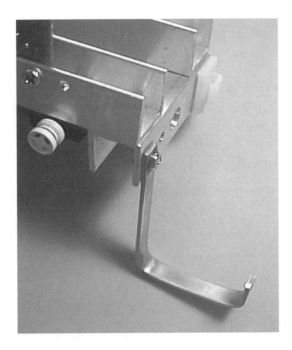

The finger attached to the x-frame reaches down beneath the right side-rail of
the main frame. Its upturned end presses against the actuator lever of MS2
when the y-frame is at the front and the x-frame is at the right. Determine the
dimensions of the finger by measurement when the x-frame is in position.

This arrangement means that the x-frame can close MS2 only when the y-frame is in its
base position. Software can take care of this requirement — first move the y-frame until
M1 closes, then move the x-frame until M2 closes.

Note that M1 and M2 are both on the main frame. Their leads run directly to the PIC1
board, from where the winch motors are controlled. If the gantry is not to become a
birds' nest of connecting wires tangling with the x- and y-frames we need to keep the
number and length of connecting wires to a minimum. This is a major reason for having
the up/down tool winch on the y-frame, rather than on the x-frame where its action takes
effect. For the same reason, MS1 and MS2 are on the main frame, only a few centimetres
away from the PIC1 processing board. Power supply for PIC2 is a problem. Do we have a
battery on the x-frame (taking up space, adding weight and increasing friction) or do we
run three wires from the main frame (tangling risk and impeding motion)? We eventually settled for the second alternative.

Tools

Which tools you provide for your gantry depends on your interest in different aspects of robotics. There is space for only one tool on the x-frame at any one time, but they are designed so that they can be quickly bolted on to the underside of the frame. Two 25 mm M3 bolts project downward at the centres of the long members of the x-frame. The selection of tools described in this part comprises:

- **Hook:** gives the *Gantry* the function of a crane. Objects to be lifted need a loop for the hook to catch.

- **Gripper:** a relatively simple one-sided design for picking up objects. Can be used for moving games pieces and for building with wooden or plastic 'bricks'.

- **Brush:** used for painting pictures.

- **Laser pointer:** Used for solving mazes, scanning pictures, playing board games.

- **Camera:** Measures brightness of reflected laser beam. Has many other applications.

Hook

The hook is made from brass rod bent into shape and attached to a pulley block. The pulley wheels used here and in other tools are included in the pulley set from Tamiya™.

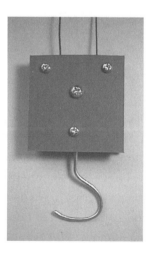

Two views of the hook, show it attached to a pulley block. The sides of the block are bolted together by two bolts at the top, with 6 mm spacers between them.

A special mechanism raises and lowers the hook. The same mechanism is used to raise and lower the gripper. The principle of this is illustrated below. The winch motor M3 is at the left end of the y-frame. From the winch the cord runs *under* the x-frame and is tied to the right end of the y-frame.

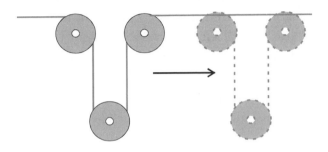

The travelling pulley mechanism. The upper two pulleys are mounted in a box on the x-frame. The lower pulley is in the pulley block that carries the hook. As the frame moves from left to right the height of the hook above ground is unchanged.

The hook is raised by winding in some of the cord. It is lowered by unwinding. But if the frame is moved without winding or unwinding the cord, the cord simply passes through the system and the hook stays at the same height. The photo below shows the mechanism in detail.

The travelling pulley mechanism with its front panel removed from the four long bolts at its corners. The rear panel is bracketed at a right angle to the tool's base panel so that it is vertical beneath the x-frame.

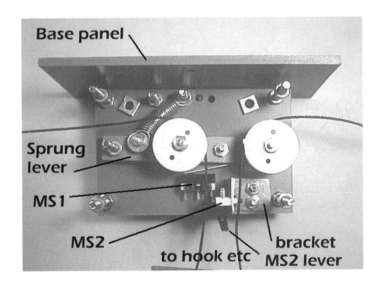

The cord is seen threaded through the mechanism as in the photo on p. 309.

The robot needs to know two facts about the hook: whether it is carrying a load and its height above the working surface. This information is obtained in a simple way by the two microswitches, labelled MS1 and MS2 in the photo. MS1 is used to detect if the hook is carrying a load. The axle of the right-hand pulley passes through holes drilled in the front and rear panels. The axle of the left-hand pulley is mounted on a lever which is pivoted at one end. This lever is held up by a spring. The upper end of the spring is bolted to one of a row of holes in the rear panel. The tension in the spring is adjustable by the choice of hole. The tension is set so that the spring supports the lever when there is no load, but allows the lever to be pulled down when a given load is present.

When the lever is pulled down it presses against the actuator lever of MS1, which closes the switch, sending a signal to the microcontroller.

MS2 is seen end-on in the figure. When the pulley block is raised as far as possible it presses up against the lever of MS2. This closes the switch, indicating that the hook has reached its upper limit. From this position a short timed burst on the winch lowers the hook to a reasonably predictable height.

There is no limit switch to indicate when the hook has dropped to ground level. However, suppose the hook is known to be carrying a load, as indicated by MS1. If the hook is then lowered, MS1 opens when the load touches the ground. The hook has reached the level at which the robot should free the hook from the load by moving the x-frame to one side. This is an instance of how software can substitute for the lack of a limit switch.

Gripper

The gripper has one fixed and one movable jaw. The fixed jaw consists of a single brass strip bent to shape and bolted to the base panel. The movable jaw is double and driven by a motor with built-in gearbox. The two arms of the movable jaw are directly attached to opposite ends of the gearbox output shaft.

The gripper is raised and lowered by the same pulley block as the hook, complete with the two limit switches.

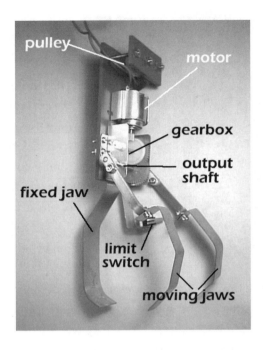

The single-sided jaw is opened and closed by the motor, through the gearbox. It is raised and lowered by the M3 winch on the y-frame.

The motor unit is compact, measuring about 40 × 64 × 20 mm. It has holes for bolting it to the PVC panel, and a clip to hold the motor. The gears are plastic and a tight fit on the shafts. There are two drive levers that are pushed on to the opposite ends of the output shaft. The motor unit is bolted to a 100 mm × 45 mm PVC base plate, which also carries the pulley and the fixed jaw.

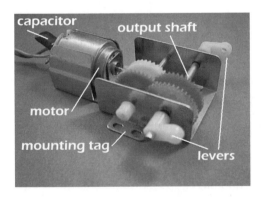

The jaw motor from CIC. The movable jaws are bolted to the two levers at each end of the output shaft. This shaft projects a few millimetres beyond the levers, fitting into one of the two holes in the base section of each movable jaw (see overleaf).

The movable jaws each have two parts. The base part is made from brass strip 5 mm wide × 60 mm long. Before bending it to shape it is drilled with two holes at one end for bolting to the drive lever and two holes at the other end for attaching the claw. The claw is made from thinner brass strip, 12 mm wide × 80 mm long. The single fixed jaw is made from brass strip 18 mm wide × 120 mm long. The overall length of the gripper is 130 mm. For some operations it might be better for the claws and fixed jaw to be shorter than shown in the photos.

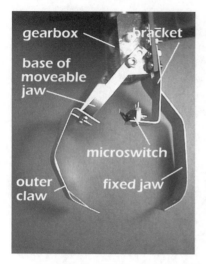

The microswitch on its bracket. The jaw base is secured to the lever by a bolt and a hole for the output shaft.

The flexibility that a gripper must have is provided by the springiness of the brass of the fixed and outer movable jaws. Even with this we need to provide feedback to the microcontroller, in the form of a third limit switch. The microswitch is mounted on a bracket cut from thin brass strip. It is positioned so that the base of one arm of the jaw closes the switch when the jaws are securely gripping an object. The position of closing can be adjusted by bending the bracket.

The limit switch provides a single reference point for the state of the jaws. Rather than install a second switch, which adds to the complexity and weight of the tool, not to mention the extra pair of wires, we rely on this single point. From this position the jaws are opened by a reasonably reproducible amount by sending the motor a pulse of controlled length.

The gripper is supported below the x-frame by the pulley mechanism used for the hook. The gripper is heavier than the hook so the anchoring point of the spring must be altered to increase the tension.

Brush

This converts our gantry into an artist. The base panel of the brush tool is bolted directly to the underside of the x-frame. The tool incorporates its own pulley mechanism for raising and lowering the brush. This is driven by the M3 winch on the y-frame.

The brush is mounted on a four-sided frame loosely bolted at the corners. This allows the brush to be moved up and down while remaining vertical. When the brush is down, in the painting position, it rests lightly on the paper. It is pulled along and paints a line as the x- and y-frames move. It is raised to stop drawing. The limit switch signals when the brush has been lifted clear of the paper and high enough to move to a paint-well to recharge the brush.

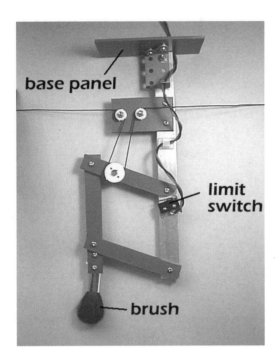

The brush tool is built from a mixture of aluminium strip, PVC strips, and Meccano parts. It has Tamiya pulley wheels, and a sawn-off make-up brush. A truly hybrid construction.

Laser

This projects an intense but very small spot of laser light on the floor beneath the x-frame (overleaf). It is used to point out objects such as playing pieces. For instance it points to the piece the *Gantry* wants to move next. Or it can trace the *Gantry*'s chosen path through a maze.

The laser is an inexpensive 'laser pointer'. It has its own battery of button cells. The pointer is held in a springy clip of the type used for holding PP3 9 V batteries. The pointer has an on-off push-button. With our pointer the laser is twisted in the clip so that the clip holds the button on.

This tool may be combined with the camera (see next section). This is directed downward to focus on the spot of laser light. The small size of the spot gives the tool high resolution when detecting objects below.

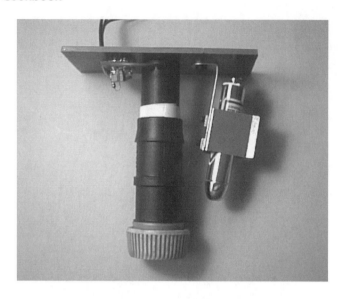

For many scanning applications the camera and laser are mounted on the same base plate.

Camera

This is a one-pixel digital camera! It has a single LDR as its sensor. Its output is an analogue voltage ranging between a few millivolts when focussed on a black object and several volts when aimed at brightly lit white objects.

Because it has a lens to focus a life-size image of the object on to the sensor, the camera has high sensitivity and a narrow field of view. These features are important when the *Gantry* is playing a board game or when registering the layout of a drawn maze. But, although the camera was designed for the *Gantry*, all of the other robots could find a use for it.

The camera lens is an inexpensive magnifier, diameter about 25 mm and focal length about 30 mm. The greater the diameter, the more sensitive the instrument. The lens is held at the end of a plastic tube that slides inside a fixed tube. The tubes are of black ABS plastic, and are plumbing items for garden reticulation systems. The retaining ring is part of a garden hose connector. The sliding tube has to stay in position after it has been focussed so it needs to be a firm fit over the fixed tube. A turn or two of PVC insulating tape around the end of the fixed tube helps to prevent slipping. Similarly, a layer of PVC tape around the focussing tube helps the retaining ring to stay in place.

The optics of the camera are as follows. A magnifying lens of focal length *f* focuses a full-size inverted image on the LDR when the LDR is 2 focal lengths behind the lens and the object is two focal lengths in front of it. For example, the lens used in the prototype camera has a focal length of 30 mm. The distance between the LDR and the lens is 60 mm. Camera is mounted so that the lens is 60 mm above the working area.

With the lens focussed at *2f,* the image of a 40 mm square on a playing board more than covers the LDR, which is only 8 mm in diameter. There is no need to aim at the exact centre of the square.

The LDR is soldered to a very small rectangle of stripboard. Two connecting wires are soldered to the board from the strip side and lead out through a hole bored in the fixed tube. The wall of the tube is 4 mm thick and the stripboard is glued to the cut end of the tube.

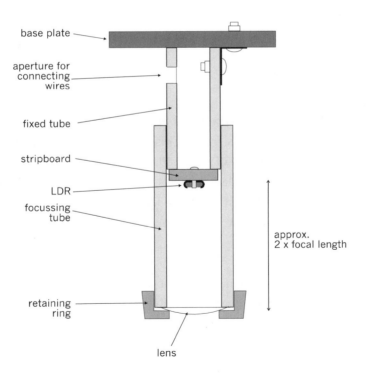

The camera, directed downward to focus an image of an object in the *Gantry's* working area.

A small stripboard bolted to the base plate carries a variable preset resistor for adjusting the output from the sensor. The circuit is on p. 75. R1 = 47 kΩ and VR1 = 470 kΩ.

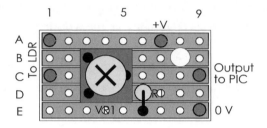

Stripboard layout of the camera board, VR1 is adjusted to suit the range of ambient light intensity.

When its output is read by the PIC comparator, the camera readily distinguishes between white and black, with or without illumination by the laser beam. If its output is fed to an AD converter it is possible to distinguish shades of grey or a limited range of colours.

Location sensors

One of the advantages of the gantry type of robot is that it is easy to keep the robot informed of the exact location of the x-frame. This is done by placing regularly spaced markers on one of the x-rails and one of the y-rails. The markers work like this. At the start of a session the x-frame is returned to its base location (front right). From then on, it keeps a count of how many markers it has passed, and the direction it is moving. From this data it can calculate its position in an imaginary square grid. The more markers, the more precisely it knows where it is.

There are two types of markers: magnetic and infrared. We used the magnetic type but the IR markers would have been almost as effective.

Magnetic markers consist of small ferrite magnets spaced evenly along two of the rails. They are detected by Hall effect devices mounted on the x- and y-frames. We used magnets measuring 10 mm square and 4 mm thick. Glue them at equal intervals to two strips of PVC board and place the strips in the channel of one of the x-rails and one of the y-rails. Make sure they all have the same pole uppermost.

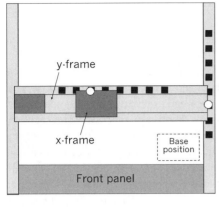

y-frame

x-frame

Base position

Front panel

○ Hall effect sensor

■ Magnet

(Left). Using markers to track the location of the x-frame and thus the x- and y- position of the tool.

(Below). Stripboard layout of the x-probe. The board is supported on two M2.5 bolts (see diagram overleaf).

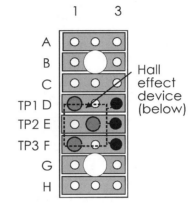

Hall effect device (below)

Two sensor probes are required and these are mounted at one end of the y-frame and on one side of the x-frame. To mount the probes, first solder each to a very small rectangle of stripboard (right).

A bracket cut from brass strip supports the x-probe so that it faces down toward the magnets, and is about 2 mm above them. Three wires (0 V, +3 V, and output) run from the probe to the PIC2 board. There they plug on to terminal pins TP1, TP4, and TP17 respectively. The height of the board is adjusted by setting the positions of the nuts that hold the board.

The y-probe is similar, but supported on a flat strip projecting from the y-frame beside the pulley. The leads run across to the x-frame and are plugged on to TP2, TP3 and TP15.

When running the leads from the probes to the PIC2 board, plan the route so that the leads will not restrict the movement of the frames. The hanging loop between the y-frame and the x-frame must not run the risk of becoming tangled with the wheels and projections such as the finger and MS2. It must not come between the end of the y-frame and MS1, so preventing the y-frame from getting to its base position.

Magnets have a north and south pole at opposite faces. The output of the sensor rises or falls when a magnet is near, depending on which pole is nearer the sensor. Usually the magnets should be glued to the strip with the same pole uppermost.

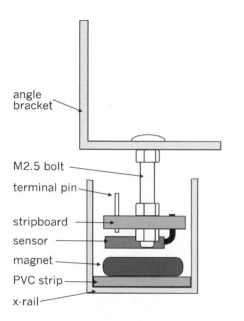

angle bracket

M2.5 bolt

terminal pin

stripboard

sensor

magnet

PVC strip

x-rail

Mounting the magnetic probe for the x-rail. The probe is seen end-on. The bracket is bolted to the side of the x-frame. The probe unit is centred by placing one or two washers between the bracket and the side of the frame. The height of the sensor is set partly by bending its terminal wires and partly by adjusting the position of the stripboard on the vertical bolts. Fine adjustment of height can also be made by inserting a strip of thin card beneath the PVC strip. The wiring to the three terminal pins passes up through a hole in the bracket and is secured to the frame by mini stick-on cable clips.

However, this polarity makes it possible to have two types of marker: north pole up and south pole up. A rise in output means one thing, a fall means another. Can this property be exploited?

IR markers are used in a similar way. The markers consist of black-painted patches on strips of white card placed inside the rails. The probes are similar to the IR probes of the *Quester* robot. The have an infrared LED to illuminate the strip and an IR photodiode to detect the reflected IR (if any). Make sure that the photodiode can not receive radiation directly from the LED or reflected from the aluminium rail. It may be necessary to push a short length of opaque plastic sleeving on to the body of the photodiode to confine it to receive only reflected radiation. Mounting the probe is the same as for the magnetic probe.

The spacing of the markers (either magnetic or reflective) depends on the intended resolution and the dimensions of the working area. Probably the best that can be achieved is an 8 × 8 grid, like a chessboard (opposite). First find the size of the working area. The size in the y-direction is the distance moved by the y-probe as the frame moves from its base position to the rear of the main frame. Measure this distance and call it Y mm. Calculate a value for X in the same way. To simplify the programming, it is better for the grid to be square and not to completely fill the working area.

In a game program, this allows the x-frame to move outside of the grid, perhaps to pick up or deposit playing pieces that are not actually in play on the grid. In painting programs, it is preferable to have the paint-pot outside of the area of the painting. The base position at the extreme front and right should also be outside the grid.

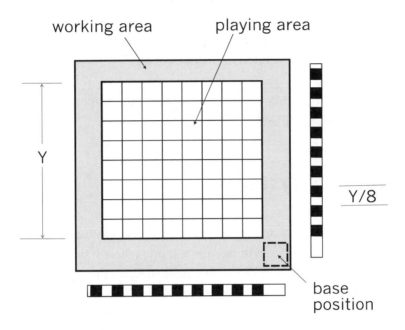

Where to place the markers. For an 8 × 8 playing area you need 9 markers for each direction.

The figure above shows that the number of markers on each side is one more than the number of squares in the grid. Starting at the base position and moving to the rear, the first marker tells the robot that the x-frame is moving into the playing area.

After this, the markers tell it as it moves from one row (or column) to the next. The last (9th) marker tells it that it is leaving the grid and must not move much further. For safety, there could be a 10th marker to stop the frame as it reaches the edge of the working area.

Decide on values of X and Y that are less than the working area. For a square grid X equals Y. Place the markers every X/8 and Y/8 along the x- and y-rails, as in the diagram.

Electronics

It is almost inevitable with any innovative project that some of the parts and components will be incompatible with some of the other parts and components. We have already come across this problem in the Mechanics section. A motor with a 2 mm output shaft has to drive a gear system based on 3 mm shafts, and sometimes our mechanism is held together by metric nuts and bolts plus a few rated in fractions of an inch.

The same happens in the electronics — for example, different components need different supply voltages. This is the first problem that must be sorted out.

Gantry and tool systems

The electronics of the *Gantry* consists of two separate but cooperating systems, each with its own PIC16F690. The gantry system moves the x-frame where it needs to go. This includes moving the y-frame. It also raises and lowers the tool. In other words, it switches the motors on or off and sets the direction in which they turn. This system also has push-buttons and a switch used by the operator, plus a pair of indicator LEDs. It gives the operator control of the robot. It is based on a PIC that we will call PIC1.

The other system, controlled by PIC2, is the tool system. This is where the important decisions are taken. It has sensors that tell it what is happening, and it decides what is to happen next. It has connections to PIC1 and signals it when to run the motors. This is a two-way connection which we will look at in more detail later.

Gantry system

In the gantry system the two main items are:

- PIC1: Operates on 2.0 V to 5.5 V DC. It needs only a few tens of milliamps to drive the LEDs.

- Motors: Three of these running on 12 V DC at up to 500 mA. Under heavier load they would need more current but 500 mA is the most they are likely to take.

The best supply for the PIC is a battery of four nickel-metal-hydride rechargeable cells. This produces 4.8 V (a little more when freshly charged). Two such batteries are convenient for running the motors. These give 9.6 V, which slightly under-runs the motors. Or we could use a 12 V battery or a mains plug-top power supply unit. The supply does not have to be stabilised, but the PSU must be capable of providing 500 mA.

Power control board

The power switching circuit is mounted on the left side panel of the gantry, near to the front. There is a 3-way screw terminal block for connections to the batteries or PSUs: motor 9.6 V or 12 V, common 0 V, and logic 4.8 V. The LEDs and switches are mounted on the side panel with a small circuit board for making the connections.

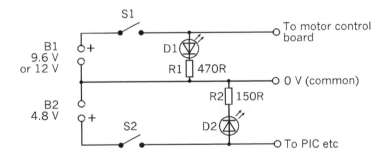

The power supplies for the PIC1. Note the common 0 V line.

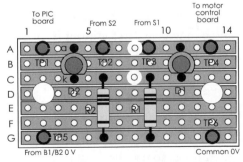

D1 and D2 are mounted on the reverse side of the board
TP2, TP3 and TP5 are 1 mm plain pins

Stripboard layout of the power switching board for the PIC1 system. This is bolted to the inside of the left side panel, low down and near the front. The bolts are 25 mm long and the LEDs at the rear of this board project through 5mm holes drilled in the panel.

This arrangement minimises the number of leads from the main frame to the y-frame. It is important to do this because stiffness of the leads is relatively greater on a small gantry like this, when compared with a full-sized industrial gantry. Also, the more dangling wires there are, the more likely one of them will get tangled with the tools and disrupt the operation of the robot.

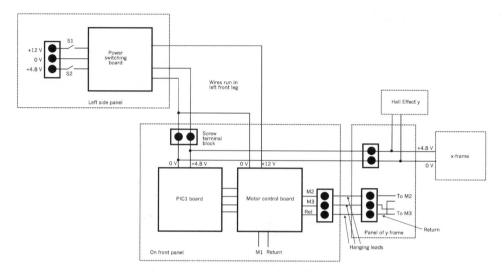

Power connections. Also shown are the control lines from the PIC1 board to the motor control board, and the output lines from there to the motors. The return line is common to all three motors.

PIC1 controller board

This carries the PIC16F690 that controls the winch motors. The schematic for the board (opposite) shows that it also has push-buttons and a toggle switch for input from the operator. It has two LEDs, one red, one green, and a bleeper for simple output signals to the operator.

The board has wired connections to the PIC2 board. This requires three wires — signal from PIC1 to PIC2, signal from PIC2 to PIC1, and 0 V — running from the front panel to the x-frame. The wires could get in the way of operations, so use the most flexible connecting wire that you can get. The wires run from the PIC1 board along the outside of the right side rail to the far right corner of the *Gantry*, then loop across to the x-frame.

Apart from the motor control relays, and the microswitches MS1 and MS2, which are connected to the board via plugs PL1-PL3, all the other items in the gantry system are on the PIC controller board, illustrated below. Input to the PIC comes from the two push-buttons S1 and S2, and the toggle switch S3. Output from the PIC goes to two LEDs, and the bleeper (AWD). There is a transistor switch Q1 useful for switching devices that take more than the 20 mA that a PIC output pin can supply.

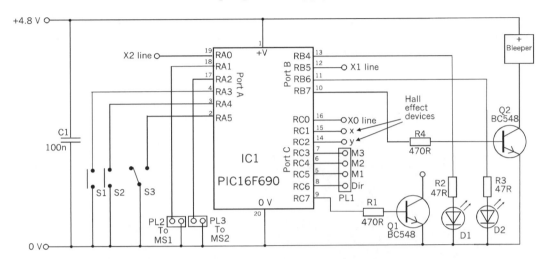

Schematic of the PIC1 controller board.

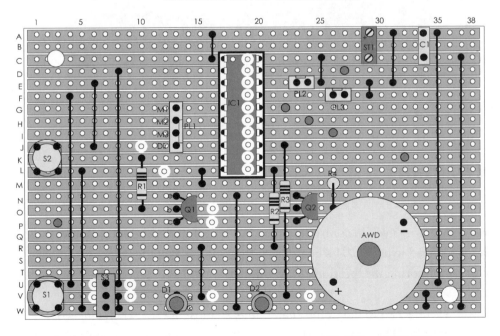

Stripboard layout of the PIC1 board. Cut strips between E23/E24 and F26/F27.

The 4.8 V supply comes from four NiMH cells, or you could use a 4.5 V or 6 V PSU. A low-powered PSU is good enough, say 200 mA or 300 mA, but the unit should have a stabilised output.

There should be no problems with the assembly of this board. You may need to modify some parts of the layout slightly to take account of differences in pin spacing for certain components, such as the push-switches and the buzzer.

There are two 3 mm holes at C3 and V36 for bolting the board to the front panel. In the prototype, it was placed at the right-hand end of the panel. We put 4 mm standoffs on the bolts. The allocation of the pins of PIC1 is listed on p. 331.

The PIC1 controller board.

Tool system

The tools and sensors used in controlling the tools are all under the command of a second PIC. This is referred to as PIC2. Its board is on the x-frame. Another function of this controller is monitoring the location of the x-frame.

Only one tool is attached to the frame at any time, so most of the I/O channels of PIC2 lead to terminal pins. There are more than enough pins to spare for the tools described on pp. 308-319, and for any that you are likely to design and build yourself.

To save space on the x-frame, the board includes a motor control circuit, used for moving the jaws of the gripper. This could be used also for driving other tools.

To save more space on the x-frame, it takes its 4.8 V power supply from the battery B2 that supplies the gantry system. This not only saves space but saves weight too. Too much weight on the x-frame increases the friction in the wheel bearings. This affects the y-frame as well. A pair of wires runs from a terminal block at the left end of the front panel. It loops to the panel of the y-frame and loops again to the x-frame. The 0 V wire is common to both systems, which is essential if signals are to be passed from one PIC to the other.

It is not feasible to provide a separate motor supply, and not worthwhile because it is needed only if you are running the gripper tool. We run it on the same supply as the PIC and accept the small risk of voltage spikes. The 3 V motor we used is known to tolerate over-running at 4.8 V but, if you prefer, choose a 6 V motor and under-run it.

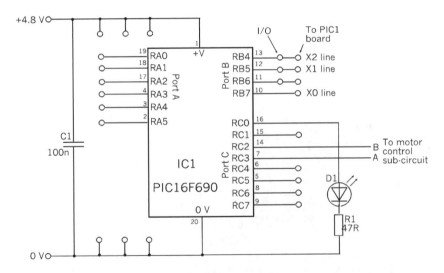

The schematic of the PIC2 board.

The schematic shows that there are many unused pins available as inputs or outputs. This gives you the maximum flexibility for designing and adding sensors and actuators to the x-frame. There are three pins allocated for communicating with the PIC1. The LED is mainly for use when testing programs. It can be temporarily programmed to light up or to flash at a certain point in the program, so confirming that PIC2 has actually got to that stage and is not hanging around at some other stage. The outputs of the magnetic sensors on the x-frame and y-frame are wired directly to PIC1.

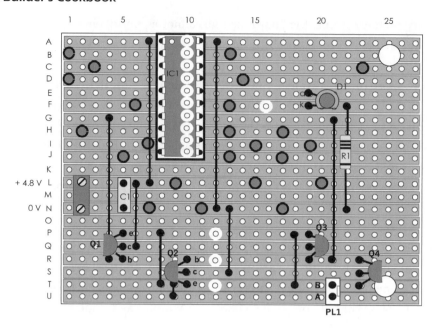

Stripboard layout of the PIC2 controller board. There is room for later additions too. There are two holes near the right edge for bolting the board to the x-frame, with its strips vertical. The empty area on the right is covered by the frame when the board is bolted in place. If you think that you may leter want to add more sub-circuits to the system, cut it to have more strips.

The PIC2 controller board mounted on one side member of the x-frame.

Motor control board

This is mounted on the front panel of the *Gantry*, and controls the three winch motors. Instead of the H-bridge, this board is based on four relays. A DPDT relay RLA1 is wired as a reversing switch. Three SPST relays RLA2-4 switch the motors on or off individually.

It receives four control inputs from PIC1, one for each relay. It has a 2-way plug output to the y-winch (M1), and a 3-way output to the x-winch (M2) and the tool winch (M3).

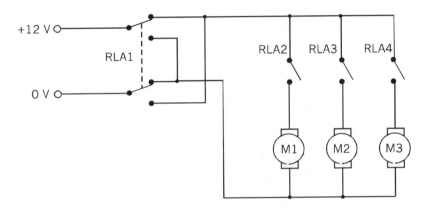

The circuit for controlling the winch motors. The motors are connected to three SPST relays, but share a common line to return the current to the DPDT reversing switch. The relays are rated to operate on 12 V, but those used in the prototype work just as well on 9.6 V from four NiMH cells.

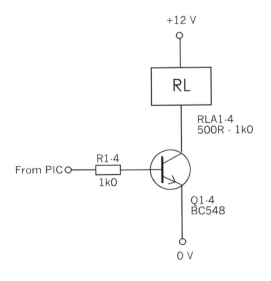

The four relays are operated by four transistor switches like the one shown in this figure. The lowest working voltage of commonly available relays is 12 V (but see above). This circuit allows the relay to be switched by a PIC operating at a lower voltage, such as 4.8 V.

The stripboard layout (below) shows the internal connections of the relays that we used. These are standard relay pinouts, so there should be no problems here, but check that your relays have the same wiring and, if necessary, alter the wiring on the board.

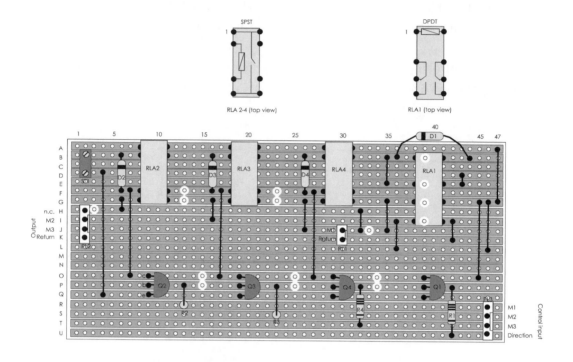

Layout of the motor control board. Strip A runs the full length of the board, with no cuts. Strip B is cut only at B39. The 12 V or 9.6 V power supply goes to the screw terminal block at B2 (+) and D2 (0 V)

Diodes protect each transistor switch from voltage spikes induced when the coils are switched off. Any small signal diode, such as a 1N4148 is suitable.

Magnetic probe circuits

Each probe (p. 317) is connected to the controller board by three leads: positive supply, 0 V and output. The positive supply and 0 V are connected to pins at the top of the PIC2 board. The output leads run direct to pins 15 (RC1) and 14 (RC2) of PIC1. These are the comparator input pins.

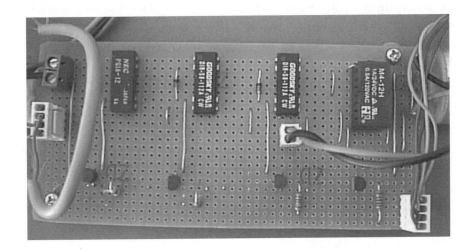

The motor control board mounted on the front panel of the *Gantry*. The y-winch motor M1 can just be seen on the right of the board.

Wiring up the tools

Several of the tools have microswitches to provide feedback to the processor. One side of the switch is connected to 0 V. The other side goes to an input channel that has been defined as a digital input with a weak pull-up. Any of the spare channels in Ports A and B can be used (see table on p. 332).

The hook is raised and lowered by the tool winch. There are two microswitches, MS1 and MS2 (p. 309). The switches operate by grounding the input. There are three leads from the pulley block to the processor board. The leads end in sockets that push on to the PCB terminal pins on the board. Each switch has a lead to the PIC. The third lead is a common ground. Use one of the 0 V pins on strip C.

The gripper has one microswitch, and this is connected in the same way as the switches on the pulley block, to one of the 0 V pins and to a spare input pin of Port A or B. This is programmed as a digital input with weak pull-up. The twin lead from the jaw motor M4 ends in a 2-way socket that goes on to the PCB header plug at V24/W24 on the processor board.

The brush has one microswitch connected to 0 V and a digital input with weak-pull-up. If you add other microswitches to the tools, there are spare inputs in Ports A and B to connect them to.

The laser has its own built-in battery. The laser tool board may also carry the camera. Its LDR light sensor operates as a voltage divider and so needs three leads: positive supply, 0 V line, and output. Use pins on strip A and C for the first two connections. The output goes to an analogue to digital converter or a comparator. This may be any of the AD input channels (AN0 to AN11) or comparator input channels (AN1 or AN5 to AN7), as indicated in the table on p. 332.

Interference

The *Gantry* has several connecting wires that are 10 cm long or longer. These wires are liable to pick up spikes and other signals radiated from other connecting wires. For example, the wires from the Hall effect sensors to the PIC1 board are about 70 cm long, and this may lead to unreliable counting of the marker magnets. This problem may not arise but, if it does, the x- and y-frames do not move the correct distances. The effect is cumulative.

The sensible solution would be to use screened cable. Unfortunately, this reduces the flexibility of the connections. It was found that the problem was eliminated by soldering 100 nF polyester capacitors to the PIC1 controller board beween the input pins and the 0 V lines. On the diagram on p. 323, the capacitors are at C32/G32 and C38/H38.

Interference may also swamp the action of the weak pull-ups on digital inputs that are made low by grounding them to the 0 V line. Examples are the inputs from the limit switches that detect when the x-frame is at its base location. Other examples are the inputs from microswitches on the tools. In such cases the input is configured not to have a weak pull-up, and instead has a hard-wired pull-up resistor connected to the positive supply line. Usually a 10 kΩ resistor is suitable, but a smaller resistance may be used. down to about 1 kΩ.

The middle diagram on p. 68 shows the circuit. The resistor can be located either on the tool or on the controller board, wherever it is convenient.

PIC1 I/O

For later reference the table below lists I/O connections to PIC1. A photocopy of this table mounted on card is handy to have on the workbench while assembling and testing the circuit modules.

Port	Name	IC pin	I/O	Connected to	0 =	1 =
A	RA0	19	I/O	X2		
	RA1	18	I	Microswitch 1	Closed	Open
	RA2	17	I	Microswitch 2	Closed	OPen
	RA3	4	I	Push button S1	Pressed	Released
	RA4	3	I	Push button S2	Pressed	Released
	RA5	2	I	Toggle switch S3	Closed	Open
B	RB4	13	O	D2 (Green LED)	Off	On
	RB5*	12	I/O	X1		
	RB6	11	O	D1 (Red LED)	Off	On
	RB7*	10	O	Bleeper	Off	On
C	RC0*	16	I/O	X0		
	RC1	15	AnI	Hall effect device (x)	Between	At marker
	RC2	14	AnI	Hall effect device (y)	Between	At marker
	RC3	7	O	Motor 3 (Tool winch)	Off	On
	RC4	6	O	Motor 2 (x-winch)	Off	On
	RC5	5	O	Motor 1 (y-winch)	Off	On
	RC6	8	O	Motors (Direction)	Decrease	Increase
	RC7	9	O	Spare trans. switch	Off	On

The entries are grouped by I/O port and list all the channels available on the PIC16F690. In the fourth column, the I or O indicates whether the channel is to be configured as an input or an output. I/O indicates that it can be either. AnI is an analogue input.

Push buttons S1 and S2: Used in games, etc. for 'Start', 'Finished my move' etc.
Switch S3: Program select, or similar functions.

X0, X1 and X2 are lines to the board on the x-frame. They control the tool or exchange signals with PIC2, if present. Lines marked * are reallocated for some of the programs on the companion site (see 'Program Notes' published on the site).

PIC2 I/O

For later reference the table below lists I/O connections to PIC2, if installed, or to terminal pins on the x-frame board. A photocopy of this table mounted on card is handy to have on the workbench while assembling and testing the circuit modules.

Port	Name	IC pin	I/O	Connected to	Comp. input	AD input
A	RA0	19	I/O	Spare	Reference	AN0
	RA1	18	I/O	Reserved for camera	Input 1 or 2	AN1
	RA2	17	I/O	Spare		AN2
	RA3	4	I	Spare (input only)		
	RA4	3	I/O	Spare		AN3
	RA5	2	I/O	Spare		
B	RB4	13	I/O	X2		AN10
	RB5	12	I	X1 (or USART input)		AN11
	RB6	11	I/O	Spare		
	RB7	10	O	X0 (or USART output)		
C	RC0	16	O	LED D1 (red)		
	RC1	15	I/O	Spare	Input 1 or 2	AN5
	RC2	14	O	Motor A		AN6
	RC3	7	O	Motor B		AN7
	RC4	6	I/O	Spare		
	RC5	5	I/O	Spare		
	RC6	8	I/O	Spare		AN8
	RC7	9	I/O	Spare		AN9

The entries are grouped by I/O port and list all the channels available on the PIC16F690. In the fourth column, the I or O indicates whether the channel is to be configured as an input or an output. I/O indicates that it can be either. Channels that can be configured as inputs to the comparators and AD converters are listed in the sixth and seventh columns.

X0, X1 and X2 are lines to the PIC1 processor board on the front panel. They send sensor information or receive signals from PIC1.

Channels in Ports A and B have programmable interrupt-on-change and pull-up facilities when acting as inputs.

Shopping list — electronic (1)

Power control board:

       R1 resistor 470R.

       R2 resistor 150R.

       D1, D2 light-emitting diode 5 mm.

       S1, S2 miniature toggle switches, SPST or SPDT.

       0.9 mm terminal pins (6 off).

       Stripboard 7 strips × 14 holes.

PIC1 Controller board:

       R1, R4 resistors 470R (2 off).

       R2, R3 resistors 47R (2 off).

       C1 capacitor, polyester, 100 nf.

       D1, D2 light-emitting diodes 5 mm (1 red, 1 green).

       Q1, Q2 npn transistors, BC548 or similar.

       AWD solid-state buzzer or siren.

       PL1 4-way PCB plug and socket.

       PL2, PL3 2-way plug and socket (2 off).

       S1, S2 push-to-make push-buttons, PCB mounting (2 off).

       S3 miniature toggle switch, SPST or SPDT.

       Screw terminals, PCB mounting.

       0.9 mm terminal pins (6 off).

       IC socket, 20-pin d.i.l., turned pin.

       Stripboard 23 strips × 38 holes.

PIC2 Controller and tool motor control board:

       R1 resistor 47R.

       C1 capacitor, polyester, 100 nf.

       D1 light-emitting diode, 5 mm.

       Q1, Q3 npn transistors, BC639 or similar.

       Q2, Q4 pnp transistors, BC640 or similar.

       PL1 2-way PCB plug and socket.

       Screw terminals, PCB mounting.

       0.9 mm terminal pins (22 off).

       IC socket, 20-pin d.i.l., turned pin.

       Stripboard 21 strips × 27 holes.

Motor control board (for winches):

       R1 - R4 resistors, 1k (4 off).

       Q1 - Q4 npn transistors, BC548 or similar (4 off).

       RLA1 DPDT relay, 12 V, miniature, PCB mounting.

       RLA2 - RLA4 SPST or SPDT relay, 12 V, miniature, PCB mounting (3 off).

[continued]

Shopping list — electronic (2)

Motor control board, (continued)
        D.i.l. IC sockets (14-pin 3 off, 16-pin 1 off).
        Screw terminal, PCB mounting.
        PL1 PCB plug and socket 2-way.
        PL2, PL3 PCB plugs and sockets 4-way (2 off).
        Stripboard 21 strips x 47 holes.

Camera board
        R1 resistor 47k.
        VR1 miniature preset potentiometer, 470 k.
        0.9 mm terminal pins (5 off).
        Stripboard 5 strips x 9 holes.

Off-board components:
        LDR1 light dependent resistor (ORP12 or similar).
        UGN3503U Hall effect magnetic sensors (2 off).
        Laser pointer.
        Ferrite magnets 10 mm X 10mm X 4 mm (12 off or more).
        Battery holder, 8 x AAA or 8 x AA, with wire or stud terminals.
        Battery holder, 4 x AAA or 4 x AA, with wire or stud terminals.
        PP3 type battery connectors (if battery box has stud terminals) (2 off).
        AAA or AA NiMH rechargeable cells (12 off) (preferred).
        Microswitches miniature (2 off for main frame, 2 off for hook, 1 for gripper, 1
                for brush).

Miscellaneous:
        Connecting wire.
        Solder.

Programming

The electronic design of the *Gantry* allows for it to be programmed in three different ways:

- Controlled by one PIC located on the front panel of the main frame. Motors M1 to M3 are controlled directly through the motor control board. Sensors and tools are located on the x-frame with wired connections to the PIC board. The PIC socket on the x-frame is not used.

- Controlled by one PIC on the x-frame, the converse of the above. Probably not so practicable.

- Controlled by two PICs, acting jointly. PIC1 is mainly concerned with the motors, and interfacing with the user. PIC2 handles the action of the sensors and tools. There are wired connections between the two PICs to coordinate their activities.

If there are numerous sensors and actuators on the x-frame, it is best for them to be processed by a PIC on the x-frame. The best plan is to adopt the third option. This not only provides increased computing power but also reduces the number of connecting wire between the x-frame to the main frame. The x-frame is able to move more freely with fewer connecting wires and there is less risk of tangling. Programming for this option is explained on pp. 155-157.

In practice, a simple system with few sensors can operate well with a limited number of wires running between the frames. This makes the first option the obvious choice. The examples in this section use six lines. These comprise two power lines and four signal lines. The signal lines are allocated different tasks depending on the tool in use and the type of activity.

Programs for the *Gantry* have the same opening stages as those for the other projects. Load the header program, which contains the directives, and add the directives for the labels used in *Gantry* programs. The listing overleaf shows these and the initial setting up of the processor.

The channels of Port A are all inputs and, since they are set as inputs by default at power-up, there is no need to do anything about configuring Port A.

```
; Labels
    delay0      equ 20h
    delay1      equ 21h
    delayn      equ 22h
    xpos        equ 23h
    ypos        equ 24h
    flags       equ 25h

    goto start
    org 04h
    goto start

start
    bcf intcon, 7       ; Disable interrupts.
    bcf status, 5       ; Bank0.
    bcf status, 6
    clrf porta
    clrf portb
    clrf portc
    bsf status, 5       ; Bank1.
    movlw 00h           ; Port B all outputs.
    movwf trisb
    movlw 06h           ; Port C all outputs
    movwf trisc         ;    except RC1 and RC2.
    bcf option_reg, 7 ; Enable weak pull-ups.
    bcf wpua, 7         ; But not RA0.
    bcf status, 5       ; Bank2.
    bsf status, 6
    movlw 60h           ; Analogue input RC1,2.
    movwf ansel
    clrf anselh
    movlw 95h           ; Set up comparators.
    movwf cm1con0
    movlw 96h
    movwf cm2con0
    movlw 0c9h          ; Ref = 0.53 x supply.
    movwf vrcon
    bcf status, 6       ; Bank0.
    bcf status, 5
```

Changing to Bank1 locations, Port B and C channels are all outputs, except for RC1 and RC2 which are to be analogue inputs for receiving signals from the two Hall effect magnetic sensors.

In Port A all channels need weak pull-ups because they are connected to switches or push-buttons that ground the channel when they are closed.

Changing to Bank2 in Port C the two input channels RC1 and RC2 are made analogue inputs.

The next five lines set up the comparators, as described on pp. 120-122. Setting CM1CON0 to 95h, sets bit <7> to enable comparator 1, bit <4> to select for non-inverted polarity of the output, and bit <2> to use the internal reference voltage. Bits <1:0> are set to 01 to assign channel RC1 (pin 15) as input. Setting CM2CON0 to 96h gives comparator 2 the same setting as comparator 1 except that its input is as RC2 (pin 14).

Before the setting for the variable reference was decided on, the PIC was removed from its socket and the 4.8 V supply turned on. A voltmeter was connected to the output from Hall sensor 1 and the 0 V line. The x-frame was moved by hand along the rail and readings taken with the sensor between the magnet markers and with it directly over the markers. With a supply voltage of 5.1 V (freshly charged), the output was 2.5 V between markers, rising to a peak of about 3 V over the markers. A sensible threshold level is 2.75 V, which is approximately 0.53 times the supply. We assume that as the supply voltage falls with the state of charge of the battery, the output and threshold voltages will fall in proportion.

A few trials with a calculator showed that the best way of setting the reference was to select for high range (see p. 121) and use the formula $V_{supply} \times (0.25 + value/32)$. With a value of 9 in bits <3:0>, the result is 0.53. Bits <7:5> are all high to use this reference for both comparators, and to select the high range. The code for VRCON is therefore c9h.

The same reference voltage must be used for both comparators but this is no problem as the sensors are identical and positioned the same distance above the magnets. However, it is a good idea to measure outputs from both sensors before deciding on the reference level.

Remember to return to Bank0 before continuing with the program.

Finding its bearings

The first thing the controller wants to know is the location of the x-frame. The x-frame may have been left anywhere at the end of the previous session. The magnetic markers are just markers, like milestones without distances carved on them. This is why there are two microswitches at the front right corner of the main frame. When the *Gantry* is first switched on the program moves the frame until both switches close. Then the PIC has a firm reference position, the **base location**.

Once it has been to the base it moves around the working area, keeping a count of the markers it passes in the x- and y-directions. From then on it always knows where it is.

The listing for moving the x-frame to base is set out below.

```
; Program begins here

gofront
      btfss porta, 1          ; MS1 closed?
      goto atfront            ; Yes.
      bsf portc, 5            ; Wind in y-winch.
      bcf portc, 6
      goto gofront            ; To check MS1 again.
atfront
      bcf portc, 5            ; Stop y-winch.
goright
      btfss porta, 2          ; MS2 closed?
      goto atbase             ; Yes.
      bsf portc, 4            ; Wind out x-winch.
      bcf portc, 6
      goto goright            ; To check MS2 again.
atbase
      bcf portc, 4            ; Stop x-winch.
      bsf portb, 4            ; Green LED on.
      movlw 05h
      call longdelay
      bcf portb, 4            ; Low to put green LED off.
```

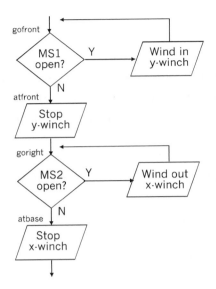

First move the y-frame toward the front until MS1 closes, then move the x-frame to the right until MS2 closes. Finally (and optionally), switch on the green LED to indicate that the x-frame is at base.

The instructions to the x-and y-winches, are given by voltage levels at RB4 (x-winch), RB5 (y-winch) and RB6 (direction):

Direction	RB4 (x)	RB5 (y)	RB6 (direction)
Left	1	X	1
Right	1	X	0
Back	X	1	1
Forward	X	1	0
Stop	0	0	X

For each direction set either RB4 (x-winch, M2) for left or right, or set RB5 (y-winch, M1) for back/forward. The setting or RB6 decides the direction. To stop the x-frame, clear either RB4 or RB5.

After moving the x-frame to base, the listing takes it into the working area to perform its tasks. These tasks use a set of subroutines for navigating the x-frame around the working area.

Moving the x-frame

One of the advantages of a gantry is that the processor always knows where the x-frame is. So one of the important programming tasks is to put it where it should be.
As explained earlier, the *Gantry* locates the x-frame on a square grid. In the prototype there are 6 rows of 6 columns, making a total of 36 locations.

There are four core subroutines for moving the x-frame. They move it one step in each direction — left, right, back, forward.

Here is the listing of *left*, the subroutine that moves the frame one step to the left:

```
left
      bsf portb, 6              ; Red LED on.
      movlw 03h
      call longdelay
      bcf portb, 6             ; Red LED off.
      bsf portc, 4             ; Winding in x-winch
      bsf portc, 6
      movlw 07h               ; 1.4 s delay.
      call longdelay          ; To clear marker.
      bsf status, 6           ; Page 2.
again1
      btfss cm1con0, 6        ; Bit set if input>0.53 x supply.
      goto again1
      bcf status, 6           ; Back to Page 0.
      btfss flags, 0
      bcf portc, 4            ; Stop.
      return
```

The flowchart is shown opposite. Before calling this subroutine set or clear bit <0> of the flags register. The eight bits of flags are used to indicate whether or not something has happened or is to happen in the future. In this program, bit <0> is clear (= 0), which indicates that the x-frame is to stop after one step. The bit is set if it is not to stop, but continue moving. This is done by using 'bcf flags, 0' or 'bsf flags, 0'.

The subroutine begins by flashing the red LED. This is not essential and can be omitted, though it is handy when testing the program.

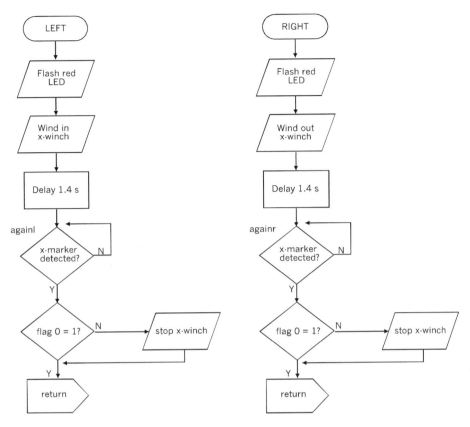

The *left* subroutine.　　　　　　The *right* subroutine.

Next, motor M2 is turned on to wind in and pull the frame to the left. There is a long delay at this stage because it is assumed that the x-frame is stationed at a marker. It must get away from that marker before it starts looking for the next one. This delay may need to be altered in length, depending on the speed of the motor and the ratio of the gearbox. Trial and error is the only practicable way to find the best length for the delay.

Now comes a two-line loop in which bit <6> of CM1CON0 is tested until it goes high.

Note the switch to Bank2 before testing CM1CON0. Note also the switch back to Bank0 after the marker has been detected.

The subroutine ends by either stopping the winch or by letting it continue running, depending on the state of flags <0>. The x-frame has moved from one marker to the next marker to the left, on the same row.

The *right* subroutine (flowchart p. 341) operates in a similar way but in the opposite direction:

```
right
      bsf portb, 6            ; Red LED on.
      movlw 03h
      call longdelay
      bcf portb, 6            ; Red LED off.
      bsf portc, 4            ; Winding out x-winch.
      bcf portc, 6
      movlw 07h               ; 1.4 s delay.
      call longdelay          ; To clear marker.
      bsf status, 6           ; Page 2.
againr
      btfss cm1con0, 6   ; Bit set if input > 0.53 x supply
      goto againr
      bcf status, 6           ; Back to Page 0.
      btfss flags, 0          ; Non-stop?
      bcf portc, 4            ; Stop.
      return
```

When typing in the listing, just copy the listing of *left* and edit it.

Moving the y-frame

This moves the x-frame too, which is the whole point of the operation. There are two subroutines, *back* and *forward* moving the y-frame by switching the y-winch, M1. Their listings are opposite. The listings and the flowcharts (p. 344) show that these have the same structure as the *left* and *right* subroutines.

Going places

The four one-step routines will take the x-frame to any part of the working area. This next routine takes it to location (0, 0), which is at the front and on the right. Starting from base where the frame was left after the routine of p. 338, it must take one step to the left followed by one toward the back.

```
back
      bsf portb, 6         ; Red LED on.
      movlw 03h
      call longdelay
      bcf portb, 6         ; Red LED off.
      bsf portc, 5         ; Winding out y-winch.
      bsf portc, 6
      movlw 07h            ; 1.4 s delay.
      call longdelay       ; To clear marker.
      bsf status, 6        ; Page 2.
againb
      btfss cm2con0, 6     ; Bit set if input > 0.53 x supply
      goto againb
      bcf status, 6        ; Back to Page 0.
      btfss flags, 0
      bcf portc, 5         ; Stop.
      return

forward
      bsf portb, 6         ; Red LED on.
      movlw 03h
      call longdelay
      bcf portb, 6         ; Red LED off.
      bsf portc, 5         ; Winding in y-winch.
      bcf portc, 6
      movlw 07h            ; 1.4 s delay.
      call longdelay       ; To clear marker.
      bsf status, 6        ; Page 2.
againf
      btfss cm2con0, 6     ; Bit set if input > 0.53 x supply
      goto againf
      bcf status, 6        ; Back to Page 0.
      btfss flags, 0
      bcf portc, 5         ; Stop.
      return
```

When the moving routines are used for a single step, only the flags <0> bit needs to be set or cleared before calling the subroutines. To get from base to location (0, 0), just clear that bit, then call left and back:

```
      bcf flags, 0         ; Enable stop in subroutines.
         call left         ; To rightmost column.
         call back         ; To front row.
```

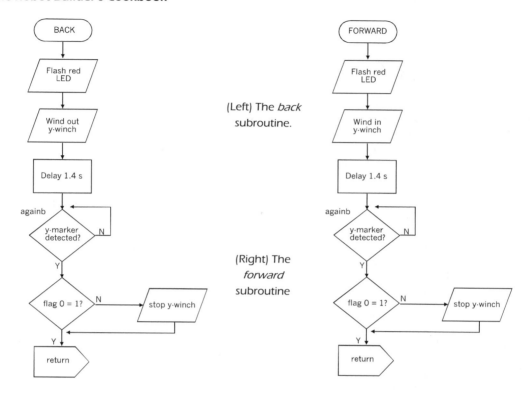

(Left) The *back* subroutine.

(Right) The *forward* subroutine

Summing up, the main program comprises:

- Directives, including labels of registers for storing data.
- Initialising.
- Moving x-frame to base.
- Moving x-frame to (0, 0).
- Subroutines: *left, right, back, forward, delay, longdelay*.
- Directive 'end'.

A useful addition to this list is *flash*, which flashes the green LEDs on and off indefinitely. This comes at the end of the main program, before the subroutines:

```
flash
      bsf portb, 4        ; Green LED on.
      call delay
      bcf portb, 4        ; Green LED off.
      call delay
   goto flash
```

The program runs straight on to *flash* after moving the x-frame to (0, 0). As the program is typed in and tested stage by stage, *flash* remains at the end of it. *Flash* is optional but it is always good to see the LED flashing to show that the controller has completed its tasks. Or, if it is flashing and the controller has *not* done everything it should have done, there is somewhere an error still to be corrected.

Moving from A to B

The routine for moving from base to (0, 0) is the simplest example of moving from A to B. It is simple because there is only one step to the left followed by one step back. Now suppose the x-frame has already been moved to (0, 0) ready to begin an action that needs it to be somewhere near the centre of the work area. It will probably need to take two or more steps to the left and back to get there. Subroutines are to be called once for each step.

Below is the listing for a routine to move the x-frame from (0, 0) to (3, 3):

```
; Move to location 3, 3.
      movlw 03h              ; Set up counters to 3.
      movwf xpos
      movwf ypos
      bsf flags, 0           ; Disable stop in subroutine.
nextleft
      call left              ; To rightmost column.
      decfsz xpos, f         ; Counting down to zero.
      goto nextleft
      bcf portc, 4           ; Stop x-winch.
nextback
      call back              ; To front row.
      decfsz ypos            ; Counting down to zero.
      goto nextback
      bcf portc, 5           ; Stop y-winch.
```

The routine has two registers, *xpos* and *ypos*, for counting the steps. To repeat the calls three times each, these registers are set to 3 and count down to zero.

Moving from C to D

The *AtoB* program is a useful one for positioning the x-frame at the beginning of a session. It assumes that it is at (0, 0) and takes it to some other location to perform its tasks. Suppose the x-frame now has to go to another position. One way of doing this is to repeat the routines for moving it to base, then to (0, 0), and eventually to the new location. This takes time and it is quicker to go directly from one location to the next, without visiting base and (0, 0). This is the function of the program named *CtoD*. It takes the x-frame from its **C**urrent location to a **D**estination location.

The routine needs six register: *xpos, ypos, xc, yc, xd*, and *yd*. We have already used *xpos* and *ypos*. The new ones are the current x, the current y, the destination x and the destination y. These registers are loaded with the required values before calling the subroutine. In this excerpt the x-frame is at (5, 4) and is to move to (3, 1). The calling routine for this is:

```
movlw 05h        ; Current x to xc.
movwf xc
movlw 04h        ; Current y to yc.
movwf yc
movlw 03h        ; Destination x to xd.
movwf xd
movlw 01h        ; Destination y to yd.
movwf yd
call xctod       ; To move it left or right.
call yctod       ; To move it back or forward.
```

The flowchart of subroutine *xctod* shows how the distance and direction are calculated from the values of the starting and finishing positions. Subtracting *xc* from *xd* gives the distance to be moved. If the result is positive, the x-frame is moved to the left, by calling the left subroutine.

The situation is slightly more complicated if *xc* is to the right of *xd*, the value in *xpos* is negative. The left and right subroutines only work with positive values. Before calling the right subroutine, the value in *xpos* must be made positive. A negative result of a subtraction causes the carry bit (STATUS<0>) to be set.

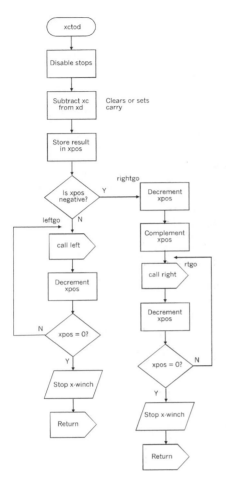

The *CtoD* routine works out the direction and distance betwwen the cuurent and destination locations,then moves the x-frame directly to it new location.

The conversion of negative to positive is done in two steps. First subtract 1 from the negative value. For example, -4 becomes -5. Then complement the byte, substituting 0s for 1s and 1s for 0s. The 'comf' instruction takes care of this and the result is the positive value. Continuing the example, -5 in hex is fah., or 11111011 in decimal. Complementing this yields 00000100, which is 4 in decimal. The *right* subroutine is then called with this value in *xpos*. The listing of *xctod* is given overleaf.

Moving in the y direction employs a similar subroutine, *yctod*. Its flowchart and listing differ from *xctod* by substituting 'y' for 'x' in all statements and labels. Also, the subroutines are called *rear* and *forward* respectively, instead of *left* and *right*.

The *ctod* routines (either one or both) can be included in the main program if they are not often used. If there is a lot of moving around it is better to list them as subroutines. Load the required values in *xc, yc, xd* and *yd* and call the two subroutines.

```
xctod
      bsf flags, 0        ; No stops.
      movf xc, w
      subwf xd,f          ; Carry=1, if result negative.
      movf xd, w
      movwf xpos
      btfsc status, 0     ; Test carry.
      goto leftgo
      goto rightgo
leftgo
      call left
      decfsz xpos, f
      goto leftgo
      bcf portc, 4        ; Stop.
      return
rightgo
      decf xpos, f
      comf xpos, f        ; Make xpos positive.
rtgo call right
      decfsz xpos, f
      goto rtgo
      bcf portc, 4        ; Stop.
return
```

Scanning

One of the more complicated operations is to scan the working area. This might be the basis of a maze solving program in which a plan of the maze is scanned into the robot's memory. Then the robot analyses the plan, solves the maze, and finally traces out the correct path by moving a pointer attached to the x-frame.

The scan has four sequences, as can be seen in the flowchart (p. 351). Starting at (0, 0) the x-frame is moved from right to left along row 0. It stops every time it reaches a marker, including the first one at (0, 0), and performs an action. In the test version of this program it simply produces a bleep. After returning from the call to *left*, it decrements *xpos*. This register is set to 5 at the start so is reduced to zero by the time it reaches (6, 0).

The next sequence returns the x-frame to the right end of the row, but without stopping or performing its action. As before, *xpos* is set to 5 at the start and reduced to zero as it reaches the right column.

In the third sequence the y-frame is moved one step back, bringing the x-frame to (0, 1). The loop then repeats from label *xloop*, taking the x-frame along the next row, stopping at each marker, and then returning non-stop to (0, 1). The *xloop* repeats, decrementing ypos each time until it has completed scanning the six rows. Note that the starting value of *ypos* is 6, but the starting value of *xpos* is 5. This is because *xpos* is decremented after calling the subroutines (left or right) but, because it is involved in the *xloop*, *ypos* is decremented before calling back. In the fourth and final sequence, the y-frame is moved forward five steps, bringing the x-frame back to (0, 0).

The scan routine can be adapted in several ways. For example when playing a game of noughts and crosses (tic-tac-toe), changing the settings of *xpos* to 2 and the settings of ypos to 3 then 2, results in the required 3 × 3 playing area. The listing continues overleaf:

```
; Scanning routine.
    movlw 06h
    movwf ypos          ; Set up y counter to 6.

; Scanning right to left.
xloop
    movlw 05h
    movwf xpos          ; Set up x counter to 5.
    bcf flags, 0        ; Stop each time.
next1
    call action         ; Demo. reads sensors etc.
    call left
    decfsz xpos, f      ; x counter = 0?
    goto next1          ; No, continue to left.
    bcf portc, 4        ; Stop.
    call action

; Returning left to right.
    movlw 05h           ; Counter to 5.
    movwf xpos
    bsf flags, 0        ; No stopping.
next2
    call right
    decfsz xpos, f      ; Until 5 markers detected.
    goto next2
    bcf portc, 4        ; Stop.
```

```
; Move y-frame one row to back.
    bcf flags, 0        ; Stopping.
    decfsz ypos, f      ; Until 5 markers detected.
    goto next3
    goto farback
next3
    call back
    goto xloop          ; To scan next row.
farback
    bcf portc, 5        ; Stop.

; Return y-frame to front row.
    movlw 05h           ; y counter to 5.
    movwf ypos
    bsf flags, 0        ; No stops.
next4
    call forward
    decfsz ypos, f
    goto next4
    bcf portc, 5        ; Final stop.
```

Operating the hook

The hook is used for picking up and putting down objects. For example, use it for moving pieces in a game. The hook is raised and lowered by motor M3 at the left end of the y-frame.

There are two microswitches to provide feedback (p. 17). M3 closes when the pulley block is fully raised and is pressing against the lever of the microswitch. This establishes that the hook is in a known position. Its function is the same as the two microswitches on the main frame, which establish that the x-frame is at its base. Once the hook is fully raised it can be lowered to any required height by switching on the motor for a given length of time. The values used in timing routines will need to be adjusted to take account of motor speed, gearing and the diameter of the bobbin.

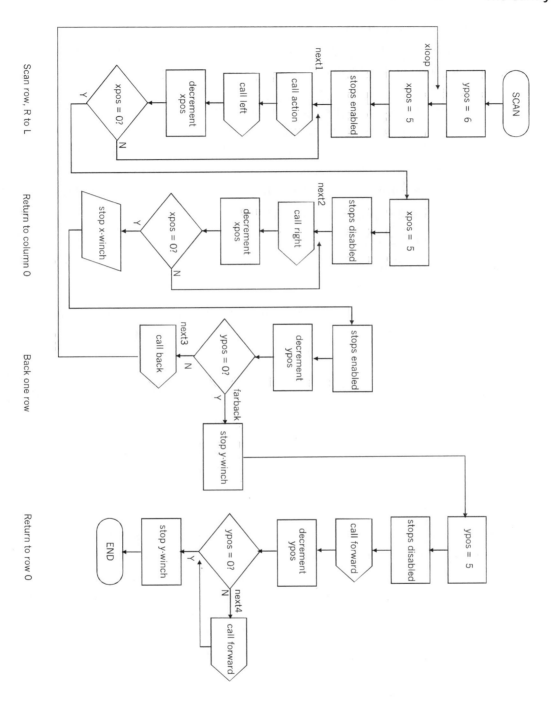

A program to scan the whole working area, and perform an action
(such as detecting an object) at each location.

The other microswitch is used to detect whether or not the hook is carrying a load. The mechanism (pp. 308-309) has a spring and the tension in this must be adjusted so that, when a load is hooked, the lever is pulled down and closes the switch.

Both switches are connected so that they ground the input pin when closed. They are wired to digital input channels with weak pull-ups. For the fully-up microswitch we used the X2 between-board link, which connects to channel RA0 (see table, p. 331). For the load-detect switch we used the X1 link, that goes to channel RB5. To make the required connection the three leads from the hook are plugged on to the PIC2 board as follows:

- 0 V to one of the 0 V pins (N7, N11, or N17).

- Fully-up switch to X2 (J13).

- Load-on switch to X1 (I15).

Each of these pins is connected directly to a second pin on the same strip. The X1 and X2 leads to the PIC1 board are plugged on to the second pin. The X0 connection is not used with the hook, but could be used for a light sensor, for example.

The listing (pp. 355-356) demonstrates how to program the hook. The x-frame moves to location (2, 3), where it picks up a load. Then it moves to location (5, 5) ad puts it down.

The load is any compact object of suitable weight (enough to close the microswitch) and with a loop or handle to pick it up by. Purists may object at having to provide a handle, but suitcases and plastic shopping bags have handles for the same purpose.

The action of this program is readily broken down into routines and subroutines. Some of these have been described already: moving to base (p. 338) and *atob* (p. 345). The *atob* routine calls *left, right, back* and *forward*, so these must be included in the listing. To assist with putting together the listing, the table on pp. 362-363 summarises all the routines and subroutines for the gantry, and the names of the subroutines that they call.

There are three routines for use with the hook:

- *raise*, which raises the hook as far as it can go.

- *lower*, which lowers the hook until the load rests on the surface.

- *updown*, which raises or lowers the hook for 0.2 s or other preset period.

Both *raise* and *lower* call on *updown*, and *updown* in turn calls on *delay*. Whether *updown* raises or lowers the hook depends on a flag that must be set before calling the subroutine. The flag is at bit <1> of the *flags* register (p. 363), where 0 = down and 1 = up.

In the original version of *raise* the hook was simply raised until MS3 closed. Unfortunately, the pressure of the hook against the pulley block caused MS4 to close and so the controller acted as if there was a load on the hook. To prevent this, a single call to *updown* lowers the hook one step after raising it.

The program begins as usual by initialising the controller. There is one difference when using the hook. In Bank 1, instead of `movlw 00h`, change this to `movlw 020h`. This sets RB5 as an input. Also add `flags EQU 25h` and `attempts EQU 26h` to the list of labels.

The next section of the program moves the x-frame to base and then to (0, 0) as listed on p. 338 and p. 343, then the program proceeds as shown in the flowchart overleaf.

 The routine for moving to (2, 3) is the same as the atob listing on p. 345, but with *xpos* set to 02h and *ypos* set to 03h.

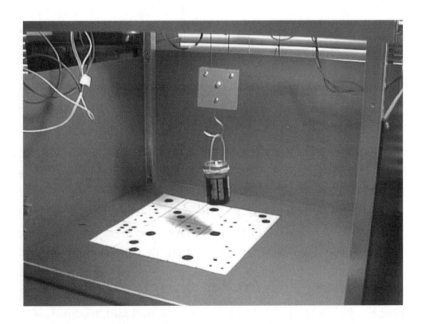

The Gantry plays a game of chance. Where will the hook deposit the load?

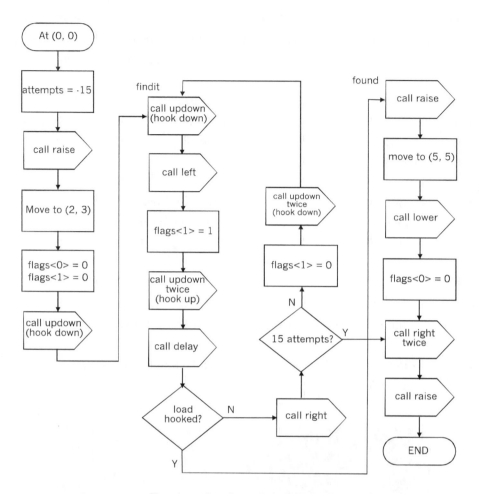

Demonstrating the action of the hook.

At this stage the program enters a routine of looking for the load. It is assumed that it will be placed just below location (2, 3), but that its height is not known. The hook is put through a sequence of moves illustrated in the diagram opposite. It tries to engage in the loop or handle on the load by moving in from the right, then moving up. If the load is not detected, it returns to the right and then drops down one step before trying again. It repeats this cycle until, after moving up, it detects that the load is engaged.

If it has tried 15 times and failed to find the load, it must be assumed that something is wrong (perhaps it has knocked the load over, or there is no load). In this case it jumps to the end of the routine, moving away, raising the hook and stopping at that point.

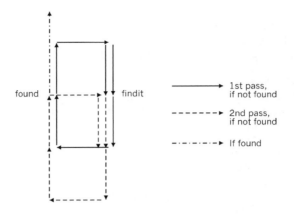

found findit

1st pass,
if not found

2nd pass,
if not found

If found

The scooping action of the hook when
searching for an object of unknown height.

```
        movlw 0fh              ; Counter to 15.
        movwf attempts
        call raise             ; Raise hook to top.

; Move to location 2, 3,

        movlw 02h              ; Set up counters to 2, 3.
        movwf xpos
        movlw 03h
        movwf ypos
        bsf flags, 0           ; Disable stop in subroutine.
nextleft
        call left              ; To rightmost column.
        decfsz xpos, f              ; Counting down to zero.
        goto nextleft
        bcf portc, 4           ; Stop x-winch.
nextback
        call back              ; To front row.
        decfsz ypos, f             ; Counting down to zero.
        goto nextback
        bcf portc, 5           ; Stop y-winch.

; Finding load.
        bcf flags, 0           ; Stop enabled.
        bcf flags, 1           ; Hook down.
        call updown
```

Controlling the hook (continued overleaf).

```
findit   call updown
     call left
     bsf flags, 1        ; Hook up.
     call updown
     call updown
     call delay          ; Time to settle.
     btfss portb, 5      ; Load detected?
     goto found
     call right
     decfsz attempts, f
     goto trymore
     goto abandon
trymore bcf flags, 1     ; Hook down.
     call updown
     call updown
     goto findit
found
     call raise

; Move to location 5, 5. (From 3, 3)

     movlw 02h           ; Set up counters to 2, 2.
     movwf xpos
     movlw 02h
     movwf ypos
     bsf flags, 0        ; Disable stop in subroutine.
nextlft
     call left           ; To rightmost column.
     decfsz xpos, f      ; Counting down to zero.
     goto nextlft
     bcf portc, 4        ; Stop x-winch.
nextbak
     call back           ; To back row.
     decfsz ypos, f      ; Counting down to zero.
     goto nextbak
     bcf portc, 5        ; Stop y-winch.

; Deposit load.
     call lower
     bcf flags, 0        ; Enable stop.
abandon  call right
     call right
     callraise

flash
     bsf portb, 4        ; Green LED on.
     call delay
     bcf portb, 4        ; Green LED off.
     call delay
     goto flash
```

The listing opposite completes the main program. The flash routine comes at the end of the main program as an indicator that the controller has completed its allotted tasks. For subroutines the program needs *delay, longdelay, left, right, back,* and *forward.*

The special subroutines for this program are as follows:

```
raise    bsf flags, 1       ; Set = raise.
         btfss porta, 0     ; MS3 closed?
         goto relax         ; Hook is at top.
         call updown
         goto raise
relax    bcf flags, 1
         call updown        ; Down a little.
         return

lower    bcf flags, 1       ; Clear = lower.
         btfsc    portb, 5  ; MS4 open? 0 = load on it.
         return             ; No load or load relieved.
         call updown
         goto lower

updown   bsf portc, 3       ; Tool winch on.
         bcf portc, 6       ; Winding in winch.
         btfss flags, 1
         bsf portc, 6       ; Or winding out.
         movlw 05h
         call longdelay
         bcf portc, 3       ; Stop tool winch.
         return

         end
```

The listing opposite shows the counters being set to move it 2, not 3 steps left. This is because the x-frame is already one step to the left when it detects and lifts the load.

The values given in this listing may need altering to take account of motor speed and other factors. Most often the length of the *delay* and *longdelay* subroutines will need increasing or decreasing, particularly in the routine for finding the load. If it is assumed that the load is never a tall one, the hook may be dropped for several seconds before it starts to search for the load. This will mean that the maximum number of attempts will be reduced.

More tasks for the hook

Combining a program similar to the above with the random number routine (p. 161) can be the basis for a game of chance. One version is being played in the photo on p. 353.

The playing area is a square of card marked out in, say, 5 rows and 5 columns of squares. The hook picks up the load from a fixed location and deposits it on one of the squares. Players win or lose according to the rules of the game, decided upon beforehand. Add excitement to the game by programming the *Gantry* to 'change its mind' a few times before eventually setting down the load.

Using a version of the *scan* routine (pp. 349-350), program the hook to sweep the area systematically for objects and place them in a neat row at the back of the working area.

Programming the brush

Like the hook, the brush is raised and lowered by motor M3, but it has only one limit switch. This is closed whan the brush is resting on the paper or is down into the paint in the well. It opens when the brush has been lifted high enough to clear the rim of the well.

The sequence of a typical painting program is: (1) move the brush to the well and dip it in the paint; (2) raise it and move it to the beginning of the stroke; (3) lower it on to the paper; (4) move it to the end of the stroke; (5) raise it. Repeat from (1) to (5) until the painting is complete.

Stages (1), (2) and (4) use the *xctod* and *yctod* routines. To program a painting, all that is needed is a look-up table of consecutive values of xd and yd. At a given stage in the sequence, xc and yc are the values held by xd and yd in the previous stage. The look-up table is constructed in the same way as a table for generating sounds (pp. 154-155). A program for painting with several colours uses two or three paint wells, plus a large well of water for washing the brush when changing colours.

Painting pictures and designs are not the only tasks for the brush. It can be used for games programs that have written output, such a *Noughts and Crosses (Tic-Tac-Toe)*.

For writing the noughts and crosses use a smaller brush. Use the *xctod* and *yctod* routines to take the brush to the selected square then call on a subroutine to draw the nought or cross inside the square.

Programming the laser

The laser can operate as a simple pointer. In a board game, for example, it points to the piece it wants to move. Its other use is for scanning the working area. If it is solving a maze, the *Gantry* uses the laser to scan the map of the maze. It scans the map into its memory, then solves the maze logically, and finally uses the laser beam to trace out its calculated path through the maze.

The laser can also operate in conjunction with the camera, as explained in the next section.

Programming the camera

The camera has a light dependent resistor, connected as a potential divider (p. 75). It can be used to measure the level of laser light reflected from a target (playing piece, playing board, maze map) in the working area. Instead of the laser, the light source may be one or more LEDs, or the ambient room lighting. When using the camera, the map or other target is best supported about 60 mm below the camera lens to bring it into focus.

For distinguishing between white (or a light tone) and black (or a dark tone) the output of the sensor circuit is sent direct to Comparator 1 of PIC2 (RA1, pin 18). CM1CON0 is set up as described on pp. 120-122. VRCON is set at a level that distinguishes a black area from a white area.

For distinguishing between a few colours (by their brightness) the output is sent to one of the 12 AD converter channels of PIC1 or PIC2. The connections from the PIC2 board to PIC1 inputs are: X0 to AN4 (pin 16), X1 to AN11 (pin 12), X2 to AN0 (pin 19).

To set up the circuit, use a testmeter to measure the output voltage under the operating conditions that are to be distinguished.

When using a comparator, use these measurements to calculate a reference voltage for VRCON that is mid-way between the two voltage levels. In the program , read CM1CON0<6>, which will be 0 for black and 1 for white.

When using an AD converter, use these measurements to calculate analogue values that discriminate between the different colours. Write an 'If A > B' branching routine based on these values to process the results of readings made at run-time.

With a little experimentation it is possible to distinguish between black, white and one or two shades of grey (or other colours such as red and green, see p. 78). This ability is useful in a game program for identifying playing pieces belonging to the two players (human and *Gantry*) and the black and white squares of the playing board.

Programming the gripper

The gripper is raised and lowered by the same mechanism as the hook, so the programming if up/down actions is the same, including the feedback from the two limit switches. In addition, the gripper motor has to be controlled by on/off and open/close signals, and there is a limit switch that closes when the jaws close.

One way of treating the programming is to make this an example of distributed processing. PIC1 controls the three winch motors and the sequencing of the operation. We use a second controller, PIC2, to control the jaw motor, under the direction of PIC1. PIC2 is located on the x-frame.

The X1 and X2 lines have the same functions as for the hook. X0 is not used for the hook but, for the gripper, its function is to coordinate control of opening and closing the jaw. The line runs from RC0 (pin 16) of the PIC1 board to RB4 (pin 13) of the PIC2 board. By configuring the channels alternately as input or output, this line is used for two-way communication between the two PICs.

The routine is for PIC1 to send a pulse along the X0 line, instructing PIC2 to open the jaw if it is closed or close it if it is open. When the state of the jaw has been altered, PIC2 sends a reply pulse to PIC1, informing it that the task is done. The flowchart opposite describes the sequence in more detail.

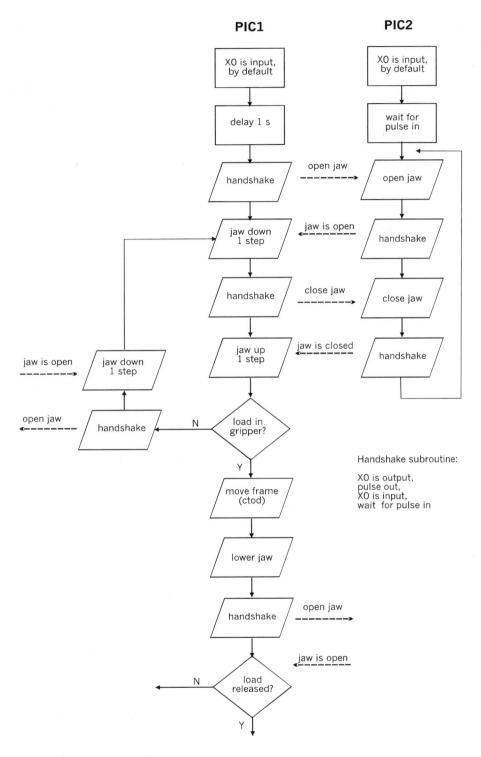

Programs with handshaking running simultaneously on two PICs.

The main program, of which only the beginning is shown in the flowchart, instructs PIC2 to open or close the jaw alternately. This is all that PIC2 has to do so, provided that all paths in PIC1's program keep to the open-close-open-close... sequence, the PIC2 program can be a loop that opens or closes the jaw, followed by sending a 'done' reply pulse every time it receives a pulse from PIC1. In between, it simply waits for the next pulse to arrive.

With additional sensors to monitor, PIC2 can have other tasks allotted to it, but the essential alternation between opening and closing the jaws must be maintained.

Gantry subroutines

This table lists the routines and subroutines used for moving the x-frame. It is a summary to use when building up gantry routines.

Name	Page	Before calling	Action	Subroutines called
left	340	flags, 0 = 0 or 1	move left 1 step	longdelay
right	342	flags, 0 = 0 or 1	move right 1 step	longdelay
back	343	flags, 0 = 0 or 1	move back 1 step	longdelay
forward	343	flags, 0 = 0 or 1	move forward 1 step	longdelay
atob (routine)	345	xpos, ypos	move no. of rows and columns set by xpos and ypos	left, right, back, forward
xctod	346-348	xc, xd, flags<0>	move from current x to destination x	left, right, back, forward, longdelay
yctod	347 (convert xctod)	yc, yd, flags<0>	move from current y to destination y	left, right, back, forward, longdelay

Name	Page	Before calling	Action	Subroutines called
scan (routine)	349-350	xpos, ypos	scan right to left, forward to back, stopping on rightward scans, return to start	left, right, back, forward
raise	357	None	raise hook to top	updown
lower	357	None	lower hook until load removed	updown
updown	357	flags <1>	raise or lower 0.2s	delay

Flags

The bits of the *flags* register have the following effects:

Bit <1>   0 = hook down; 1 = hook up.
Bit <0>   0 = stop at end of left/right/back/forward subroutines; 1 = no stop.